Mamadou Seydou Diallo

Climate Change in Sub-Saharan Africa: a spectrum of disasters?

Mamadou Seydou Diallo

Climate Change in Sub-Saharan Africa: a spectrum of disasters?

Perspectives and Challenges

LAP LAMBERT Academic Publishing

Impressum / Imprint

Bibliografische Information der Deutschen Nationalbibliothek: Die Deutsche Nationalbibliothek verzeichnet diese Publikation in der Deutschen Nationalbibliografie; detaillierte bibliografische Daten sind im Internet über http://dnb.d-nb.de abrufbar.
Alle in diesem Buch genannten Marken und Produktnamen unterliegen warenzeichen-, marken- oder patentrechtlichem Schutz bzw. sind Warenzeichen oder eingetragene Warenzeichen der jeweiligen Inhaber. Die Wiedergabe von Marken, Produktnamen, Gebrauchsnamen, Handelsnamen, Warenbezeichnungen u.s.w. in diesem Werk berechtigt auch ohne besondere Kennzeichnung nicht zu der Annahme, dass solche Namen im Sinne der Warenzeichen- und Markenschutzgesetzgebung als frei zu betrachten wären und daher von jedermann benutzt werden dürften.

Bibliographic information published by the Deutsche Nationalbibliothek: The Deutsche Nationalbibliothek lists this publication in the Deutsche Nationalbibliografie; detailed bibliographic data are available in the Internet at http://dnb.d-nb.de.
Any brand names and product names mentioned in this book are subject to trademark, brand or patent protection and are trademarks or registered trademarks of their respective holders. The use of brand names, product names, common names, trade names, product descriptions etc. even without a particular marking in this works is in no way to be construed to mean that such names may be regarded as unrestricted in respect of trademark and brand protection legislation and could thus be used by anyone.

Coverbild / Cover image: www.ingimage.com

Verlag / Publisher:
LAP LAMBERT Academic Publishing
ist ein Imprint der / is a trademark of
OmniScriptum GmbH & Co. KG
Heinrich-Böcking-Str. 6-8, 66121 Saarbrücken, Deutschland / Germany
Email: info@lap-publishing.com

Herstellung: siehe letzte Seite /
Printed at: see last page
ISBN: 978-3-659-48468-1

Zugl. / Approved by: Brussels, Vrije Universiteit Brussel,, Diss., 2000

TABLE OF CONTENTS

LIST OF FIGURES

LIST OF TABLES

LIST OF BOXES

Abbreviations

AIJ	Activities Implemented Jointly
AOSIS	Alliance of Small Island States
CDM	Clean Development Mechanism
CEIT	Countries with Economies in Transition
CERs	Certified Emission Reductions
CFCs	Chlorofluorocarbons
CH_4	Methane
CO_2	Carbon Dioxide
COP	Conference of the Parties
ENSO	El Nino/Southern Oscillation
ET	Emission Trading
FAO	united Nations Food and Agriculture Organization
GDP	Gross Domestic Product
GEF	Global Environmental Facility
GHG	Greenhouse Gas
GWh	Giga Watt Hour
ICSU	International Council for Scientific Unions
INC	Intergovernmental Negotiating Committee
IPCC	Intergovernemental Panel on Climate Change
JI	Joint Implementation
MEPN	Ministère de l'Environnement et de la Protection de la Nature
N_2O	Nitrous Oxide
OCDE	Organization for Economic Co-operation and Development
OPEC	Organization of Petroleum Exporting Countries
QELROs	Quantified Emission Limitation and Reduction Objectives
SBI	Subsidiary Body for Implementation
SBSTA	Subsidiary Body for Scientific and Technological Advice
UNCED	United Nations Conference on Environment and Development

UNEP	United Nations Environment Program
UNFCCC	United Nations Framework Conventions on Climate Change
WCRP	World Climate Research Program
WHO	World Health Organization
WMO	World Meteorogical Organization
WRI	World Resources Institute
WWF	World Wide Fund for Nature

Chapter One
Introduction

1.1 Background Information

The United Nations Conference on Environment and Development (UNCED) held in Rio de Janeiro (Brazil) in June 1992 provided the international community with the opportunity to declare that it is aware of the numerous threats which are hanging over the global equilibrium of the Earth. The huge increase of the world population combined with the degradation of the environment through deforestation, expansion of deserts, pollution, anarchic urbanisation, excessive and unconsidered consumption of the natural resources inclined a number of participants to adopt an alarmist tone and to call for changes in the direction of development. One of the issues that generated the biggest concern was the global warming and the subsequent prospects of *climate change*.

The industrial revolution and the outstanding technological advancements achieved during the last two centuries have brought about changes in the chemical composition of the atmosphere, precisely by enhancing the concentration of greenhouse gases such as carbon dioxide, methane and nitrous oxide. These trace gases, particularly the carbon dioxide, seem to have exceeded the levels of concentration needed to establish the basic greenhouse effect necessary to maintain optimal temperatures at the surface of the earth. Too large amounts of these gases are being released by the human activities and are accumulating in the atmosphere. With their heat-trapping properties, they are gradually increasing the average global temperature. This phenomenon is expected by scientists to lead to some changes in the climate system at global and regional scales. A cataclysmic vision of rising sea and devastating drought is often used to illustrate the direct virtual effects of *climate change*. Very likely, these potential effects, if they occur, would not directly affect all parts of the world with the same intensity and even may not at all concern certain regions where climate change may generate beneficial effects. So, the

1

vulnerability[1] of the countries is not a homogeneous fact. Rightfully, the industrialised countries, which are considered as the largest contributors of carbon dioxide emissions, are compelled by the United Nations Framework Convention on Climate Change (UNFCCC) to reduce these emissions in order to stabilise the level of concentration in the atmosphere and to curb global warming. In addition to being responsible for the occurrence of global warming, the developed countries are supposed to possess the economic power and the technical and institutional capabilities to overcome the problems that might threaten their security and their future. The Southern countries especially the African countries are for their part in the difficult position. They have historically contributed little or nothing to the problem of global warming and yet, they are more vulnerable to the worst of the potential effects. The images of a bruised Sahel regularly suffering from drought and hunger since the seventies and those of Bangladesh in prey to a destructive flooding in the early nineties are still persistent in everybody's mind. More recently, Mozambique has been victim of torrential rains and one third of the country completely covered by the flood. Hundreds of people died, the agricultural fields were destroyed, parts of the livestock were decimated and now the economic potential seems to be considerably lessened. The reports broadcast by all the televisions of the world showing people of all ages perched on trees for days and waiting desperately for assistance that the country itself could not bring have moved many persons to tears. Sad and impressed by the extent of the damages, certain African leaders, in particular heads of States, asked for the cancellation of the debts of the country. Such events are maybe a foretaste that foreshadows the coming disasters. For this reason, they justify people's legitimate apprehensions with regard to some dark predictions associated with global warming and climate change. Certain optimistic persons could urge that there is no need to get upset and to be agitated because of the lack of certainty in the patterns of change predicted by the scientists. But in circumstances like those created by the perspective of climate change, do the international community and the different national governments, especially those in Sub-Saharan Africa, have the right to adopt a wait-and-see policy? Isn't it more reasonable and responsible to anticipate on the coming events by taking into

---

[1] Vulnerability defines the extent to which climate change may damage or harm a system. It depends not only on a system's sensitivity but also on its ability to adapt to new climatic conditions (IPCC 1995)

serious consideration the scientists' predictions? It would be certainly interesting to try to answer these questions in the context of Sub-Saharan Africa characterised by current unfavourable natural conditions and alarming socio-economic situation.

1.2 Relevance to Human Ecology

Climatic variability is a natural feature of the climate system. Phenomenon such as severe volcanic eruptions can induce disturbances in the equilibrium state of the climatic system (Bolle 1998). However, since in such cases the equilibrium state is relatively rapidly recovered by the climate itself, the natural climate variability cannot inspire a huge concern. What is generating an enormous anxiety is the present situation characterised by the possibility of an anthropogenic climate change. As mentioned before, the human activities are causing changes in the atmosphere by increasing the amount of greenhouse gases. Those changes are expected to have significant adverse impacts on the natural and socio-economic environment of many regions of the world. Hence, decisions and actions are required from both scientists and policy makers world-wide, at the international level, what is already being done within the framework of the UNFCCC, as well as the national levels, in order to take up the coming challenges. Which types of problems climate change could impose to African countries in the coming decades? Should they passively wait for the misfortune to crush them or should they from early times set the means to overcome the difficulties? Which policies should they adopt to minimise the potential impacts of Climate Change on their natural environment and on their perspectives of socio-economic development? These are the questions raised by this research project. Naturally, such questions cannot be ignored by a discipline like Human Ecology, which is trying to have through a holistic approach a broad understanding of the problems of our time, by the study of human cultures and their role in the interactions between man and his environment.

As known, the climate change issue is mainly based on the assumption that man and the model of development followed since the industrial revolution are likely to bring about dangerous disturbances in the climate system. It seems that the human societies are about to disrupt the natural equilibrium existing in the climate system mainly through the

3

industrial activities. For this reason, it is feared that the coming changes will have diverse effects on the environment and directly or indirectly affect all aspects of people's life. Consequently, it is vital to find the means to avert these perspectives and to anticipate on the presumed adverse impacts of climate change. The challenge is to find the adequate responses to climate change, which is a threat with several dimensions and to enhance the ability of the world to follow sustainable development paths. Considered in the light of this challenge, Human Ecology appears as a discipline that can legitimately be involved in such research project.

1.3 Objectives

This research project is a prospective study motivated by the fact that many African countries could be severely affected by the impacts of future climate change whereas they possess very limited technical, financial and management resources to tackle the effects by the design and the implementation of adaptation measures. Although it is centred on the themes of global warming and climate change, it is not an exercise of climatology discussing issues like the atmospheric general circulation system or the variability and geographical distribution of insolation, etc. Using the results of studies carried out by different institutions and persons on the regional impacts of climate change, it seeks simply to stress the extreme vulnerability of countries that are often disarmed and powerless in case of disasters. It would also indicate the options that should be privileged to address the coming problems.

1.4 Working Hypothesis

The study is based on two working hypotheses:

- The first is that climate change is likely to compromise and to delay the perspectives of socio-economic development in Sub-Saharan Africa. The environmental stresses and other extreme events such as drought, desertification, floods that certain African countries are currently facing and their associated damages in particular on the socio-economic activities (agriculture, tourism, fisheries, water and energy supply) and on ecosystems and human settlements, may be seriously exacerbated.

- The second is that orienting the development of Sub-Saharan Africa along a sustainable path is the only reasonable way to reduce its vulnerability and to improve its resilience to eventual climate changes.

1.5 Methodology

The research methodology consists of an analysis of literature, mainly derived from:

- The publications of scientists and institutions such as the Intergovernmental Panel on Climate Change (IPCC) working on the different aspects of the climate change issue;

- Some national reports in particular the national communications that the country Parties to the United Nations Framework Convention Climate Change have to submit regularly to the Conferences of Parties (COP);

- The resources on climate change provided by Internet.

1.6 Limitations (warning)

As stated above, this work is not an exercise for climatologists. It is not designed to discuss the scientific validity of the predictions made by the researchers. Climate Change impact assessment is a young discipline that is still in construction. The scientists involved in its use are the first to declare that the understanding of future climate change and its impacts is still surrounded by some uncertainties. But there is more and more convergence in the views expressed by the scientific community especially within the framework of the Intergovernmental Panel on Climate Change (IPCC). In these conditions, the scientists' opinion can reasonably be presumed credible and considered as a broad indication of what might happen in the coming decades. So, this work proceeds from the refusal to be paralysed by the uncertainties and the determination to make a modest contribution to the need to deal with the perspectives of climate change in Africa.

Chapter Two
The Scientific Background on Climate Change

2.1 The Greenhouse Effect

Considered at the global, regional or local scale, the climate is generally regarded as a system involving different elements, particularly the atmosphere, the oceans and the biota (land surfaces and the vegetation, snow and ice cover). Its different patterns, which give each locale its specific character, result from the combination of complex sets of interactions occurring within the atmosphere and between the atmosphere and the other elements of the system. Operating like a machine, the system is fuelled by the sun. The role played by the solar radiation is essential because the distribution of the sunlight reaching the earth and the amount of heat emitted from it determine the character of the global climate. Naturally, the magnitude of the energy flows is not equable over the time. Different forcing factors such as the variations of cloud cover, the orbital changes of the earth around the sun, the state of the land surface of the continents and the evolution of the chemical composition of the atmosphere can affect the balance between the incoming and the outgoing radiation. Normally, the balance of the earth's energy budget is established by a phenomenon commonly called greenhouse effect. This is now abundantly evoked in the scientific literature dealing with the perspectives of climate change. It has even come into widespread use in the popular language. However it is sometimes mentioned in terms that give the feeling it is originally a completely man-made phenomenon. That is not correct. The misuse or misunderstanding of greenhouse effect is probably rising from the fact that it is based on principles similar to those prevailing in a heat-trapping glass installed in a horticultural greenhouse. But here, instead of glass or human construction, we are dealing with heat-trapping gases in a process that is entirely and perfectly natural. The atmosphere which is mainly made of nitrogen, oxygen and argon, contains also in relative small quantities other gases such as water vapour, carbon dioxide, methane and nitrous oxide called for this reason trace gases. These gases have the singular property of absorbing the thermal infrared radiation emitted by the earth's surface whereas they are pervious to the sunlight, composed of short-wave radiation, falling on the earth (Mintzer 1992). By doing so, they naturally warm the atmosphere which in turn, re-radiates heat in all directions and especially downward to the Earth, heating the low atmosphere and the water and land below. Generated by the layer of heat-trapping gases present in the atmosphere, this additional heating forms what is widely known under the name of greenhouse effect. However, all the trace gases do not have the same contribution to its occurrence because of differences in radiation absorption, atmospheric lifetime and concentration (Le Treut 1997). Water vapour and carbon oxide are the most important since they respectively account for 65% and 25% of the natural greenhouse effect. As a whole, the greenhouse gases produce an essential phenomenon that makes life possible on the Earth. Without it, the current average temperature of the Earth that is 15° C would be lower, about –18°C and so, water would not have existed as liquid to enable the biological evolution of life. Furthermore, our planet would have had a destiny similar to those of Mars and Venus where the *greenhouse effect* is respectively too weak (average temperature -47° C) and too strong (average temperature 477° C) and represents a serious constraint to the development of life (Whyte 1995).

Figure 2.1: The Greenhouse Effect

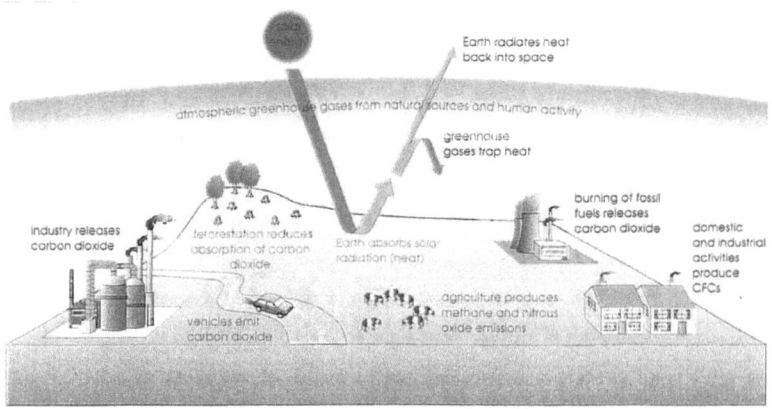

Source: UNEP 1993

7

Unfortunately, the natural greenhouse effect is now going beyond an appropriate level. It is becoming less and less natural because of an unconsidered increase of the atmospheric concentration of many heat-trapping gases generated by human activities, especially from the industrial revolution. This is why the greenhouse effect is no longer regarded as a simple natural component of the Earth's geophysical balance. On the contrary, it evokes in people's mind an abnormal phenomenon associated with the enhancement of the warming effect, which results from a growing accumulation in the atmosphere of greenhouse gases released mainly by industrial and agricultural activities. As a result, the annual average temperature of the earth is engaged in an upward trend. The mean global temperature has increased by up to 0.6° C since the mid-nineteenth century (Whyte 1995). The warmest years since historical records have been kept have occurred recently during the past two decades with 1990, 1995, 1997 and 1998 among the warmest (Justus 2000). So, it is no longer possible to have doubt about the reality of a warming climate.

Figure 2.2: Evolution of the Global Average Surface Temperature, 1860-1998

Note: While the bars represent the annual temperatures the curve indicates the five-year average.

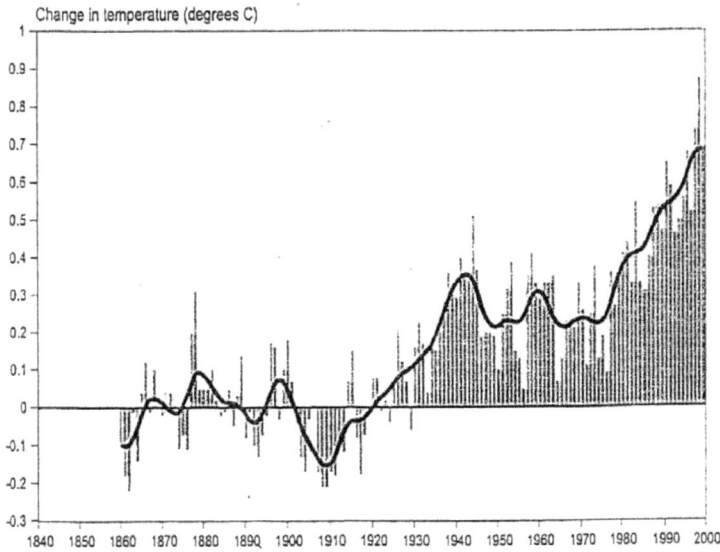

Source: Hadley Centre 1998 cited in Oberthur and Ott 1999

8

2.2 The Greenhouse Gases and the Human Activities

It is now widely accepted by the scientific community that human activities are changing the composition of the atmosphere by increasing the concentrations of the so-called greenhouse gases. This is alarming enough because it is occurring at an unprecedented speed and the trends seem to indicate a sustained process. If the trends persist, the Earth's climate system is likely to experience serious disruptions. As mentioned above, the incriminated gases, identified by the scientists as the most important greenhouse gases whose atmospheric concentrations are influenced by man, are carbon dioxide, methane, nitrous oxide, and the halocarbons, particularly the chlorofluorocarbons (CFCs) and ozone. However, though they have the common property to be heat trapping, they are generally produced from different sources with various levels of human implication. Water vapour is not taken into consideration in spite its status of largest contributor to the natural greenhouse effect because the focus is on the contributors to the "enhanced greenhouse effect".

2.2.1 Carbon dioxide

Primarily produced through metabolism (men and animals metabolising the biomass) and natural decomposition by micro-organisms, the carbon dioxide present in the atmosphere is theoretically maintained at a stable level by the ability of vegetation and plankton to catch it and to stock the carbon as source of energy. Moreover, the use of biomass as combustible, which naturally releases carbon dioxide in the atmosphere, could not disturb this balance as long as the trees cut to produce woodfuel are replaced by new ones at the same rate.

But, this situation of relative stability is far from being embodied in the real world because the carbon dioxide concentration in the atmosphere did not stop increasing since the beginning of the industrial era. It has increased by 25% to 30% passing from 265 and 280 molecules per million other molecules to about 353 ppm in the present time (Nilsson 1992). The analysis of air bubbles contained in ice cores collected from the poles has enabled the scientists to make this observation. It seems that is the result of human activity, which has generated through the use of fossil fuels such as coal, natural gas and oil and deforestation and land use change over three hundred billions of tons of carbon dioxide during the last two centuries. The amount of carbon by which the current releases exceed the absorption and sequestration by natural processes into the terrestrial biota and into the oceans is 6 billion tons per year. This amount accumulates in the atmosphere, increasing ceaseless the concentration of carbon dioxide (Nilsson 1992). Over the last 200 years, carbon dioxide has contributed for more than 60% to the enhanced greenhouse effect. Its contribution to the natural greenhouse effect is estimated to only account for 25% far behind the water vapour (Whyte 1995). The role of fossil fuels in the emissions of carbon dioxide is considered to be twice more important than that of deforestation and land use change.

2.2.2 Methane

The metabolic process through which anaerobic bacteria extract energy from dead organic matter lead to an important production of methane. This bacterial activity occurs mainly in swamps, in paddy fields, in landfills and curiously too, in the guts of grazing animals and termites, which teem with methane producing bacteria. Besides, as a natural gas subject to some exploitation, methane is also released by leakage during its transportation in pipelines. It can also escape to the atmosphere from coal mining or from an incomplete combustion of biomass and fossil fuels.

However, it is not an easy task to formulate good estimates of methane emissions from these different sources. Estimates of emissions from paddy fields are made difficult by the fact that many factors intervene such as the extent of the land surface, the density of the rice plants, the use of fertilisers, the water management, etc. On the other hand, quantification from the grazing herbivores is more or less accepted even though certain scepticism exists inspired by the fact that one cannot reasonably compare the skinny animals of the poor countries with the well-fed ones of the developed world.

But there is no doubt that the current atmospheric concentration of methane is greater than at any time in the past 160,000 years. Since the mid of the 70s, from which started detailed measurements of the amount of methane present in the atmosphere, its concentration is increasing by more than 1% per year. Moreover, from the analysis of air bubbles trapped in ice sheets, it has been demonstrated that methane concentration has increased tremendously during the past 300 years (Nilsson 1992). There is a strong temptation to impute this increase to human activities, especially rice cultivation and animal husbandry. There are more and more cattle and more and more harvested rice. Besides, one should not neglect the exploitation of natural gas that has expanded exponentially. Because of the trends of population growth, the emissions of methane are also expected to increase. That perspective can be positively or negatively influenced by climate change itself. Less methane is produced in case of reduction of soil moisture whereas there will be more if vast areas of permafrost thaw (Nilsson 1992).

The methane present in the atmosphere is generally broken down to water and carbon dioxide. But before this atmospheric oxidation, which leads to the generation of another greenhouse gas, the methane may stay there for about 8 to 11 years. It has a radiative forcing higher than that of carbon dioxide, 21 times by molecule and 58 times by weights (Whyte 1995). To maintain the atmospheric concentration of methane at current level, the global man-made emissions should be cut by 15 to 20% (Nilsson 1992).

2.2.3 Nitrous Oxide

There are various sources of nitrous oxide emissions; some are natural whereas others are directly related to human activities. The soils, especially those located in tropical forest regions, are considered to be the most important source of the nitrous oxide released into the atmosphere. Probably, the nitrous oxide generated from that source derives from processes of denitrification occurring in aerobic conditions. When the forests are cleared for the expansion of agricultural fields, the nitrogen bound in the soil is naturally released to the atmosphere. This natural production of nitrous oxide from the soil is considerably increased in cultivated grasslands by the use of nitrogen fertilisers. The atmospheric reservoir is also provided with nitrous oxide by the oceans, precisely from upwelling areas where deep water moves to the surface and in less important proportions, by the burning of fossil fuels and biomass.

According to the results of ice core bubbles analysis, the atmospheric concentration of nitrous oxide during the pre-industrial period was relatively stable at 285 ppb. Naturally, that situation has evolved establishing the current atmospheric concentration at a higher level, about 310 ppb corresponding to an atmospheric reservoir of 1500 million tonnes (Cowie 1998). It seems that the industrial production of nylon in which is used nitric acid, is particularly involved in that increase of nitrous oxide atmospheric concentration. Over the last 200 years, its contribution to the enhanced greenhouse effect is only 4%, which are relatively low compared with the 17% of methane and 61% of carbon dioxide (Whyte 1995).

As a stable chemical, nitrous oxide has a lifetime relatively long. It can remain in the atmosphere for some 150 to 170 years. However, it is decomposed in the upper atmosphere through oxidation to produce oxygen, nitrogen and nitric oxide. The latter is unfortunately involved in the destruction of the ozone layer (Nilsson 1992). Soils and aquatic systems are also regarded as minor sinks. But that point is not cleared up by any serious assessment to determine their exact role. In the light of all these considerations, the reduction of nitrous oxide emissions will not be only useful for tackling the global warming issue; it will also be of interest for the protection of the ozone layer.

Figure 2.3: Increase in atmospheric concentrations of Carbon Dioxide, Methane and Nitrous Oxide since 1750.

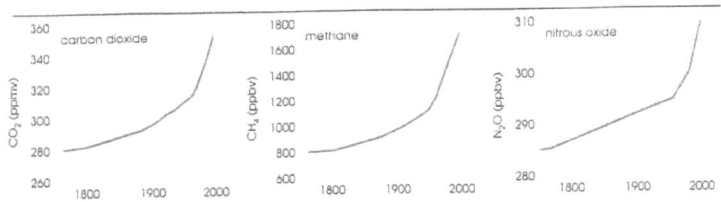

Source: UNEP 1993

2.2.4 The halocarbons

This group of chemicals is considered to be responsible for the third most important contribution to global warming. Among the greenhouse gases, they have a particular status since they are almost entirely man-made. Excepted a few, such as methyl bromide and methyl chloride that are mainly released from the oceans, most halocarbons are produced by human activities and used in different fields as refrigerating fluids, solvents, fire retardant, propellant gas, etc.

Besides, the halocarbons are involved in two major global environmental issues: the depletion of the ozone layer and the global warming. Among the large public, they are first known for their destructive potential over the stratospheric ozone and for their current regulation under the Montreal protocol. Efforts are made within this international agreement in order to phase out by the current year the main halocarbons that deplete the ozone. Their capacity to act as greenhouse-forcing agents is also widely recognised now. This capacity is even striking because it is higher than that of any other greenhouse gas. For instance, CFC_{12} has a heat-adding impact 16,000 times higher by molecule and 5,200 times higher by weight than carbon dioxide. In addition, the Chlorofluorocarbons (CFCs) absorb infrared radiations that cannot be absorbed by other greenhouse gases because of the specific wavebands of those radiations.

The high stability of the CFCs added to their properties evoked above confers on them a considerable harmfulness for both the ozone layer and the warming effect. Since the first measures carried out in the beginning of the 70s, some halocarbons have known among the greenhouse gases the most important increase in terms of atmospheric concentration, about 500% (Cowie 1998). This is naturally alarming and despite the efforts made through the implementation of the Montreal protocol that are slowing their atmospheric concentration, the halocarbons are likely to keep their important role in the generation of the enhanced greenhouse effect.

2.2.5 Ozone

Operating like an atmospheric shield that protects the Earth and its inhabitants against the Ultra-Violet light, ozone is well placed on the international political agenda because of its alarming depletion and the efforts made within the Montreal Protocol to phase out the chemicals that deplete it. Present both in the stratosphere and in the troposphere, its concentration is higher in the former, about 90%, where it contributes to regulate the climate by absorbing parts of the incoming and outgoing energy heat. In the other hand, the presence of ozone in the troposphere is associated with the enhanced greenhouse effect. But, a lack of data prevents a quantification of its contribution to the warming effect and for this reason, many scientists prefer to ignore its role as climate-forcing gas. However, the production of tropospheric ozone is known to proceed from photochemical reactions in the presence of methane, other hydrocarbons, carbon monoxide and nitrogen oxides, all generated by combustion processes involving fossil fuels. It is also known that its lifetime is relatively short, about a few weeks, before breaking down into oxygen atoms under the action of ultraviolet radiation. Besides, its distribution in the atmosphere is not homogeneous. The concentrations are higher over the industrialised areas and in this regard, it seems that Europe is experiencing an increase of 1-2% per year and the concentrations have doubled since the 50s (Whyte 1995).

Table 2.1: Principal Greenhouse Gases (Ozone not included because of lack of data)

Greenhouse gases	CO_2	CH_4	N_2O	CFC11	CFC12
Pre-industrial atmospheric concentration	280ppmv	0.8ppmv	288ppbv	0	0
Concentration in 1994	358ppmv	1.72ppm	312ppbv	268pptv	484pptv
Annual increase in the atmosphere	1.8ppmv	0.01ppm	0.8ppbv	–	–
Atmospheric lifetime (years)	50-200	12	120	50	130

Source: Cowie 1998

ppmv = parts per million by volume; ppbv = parts per billion by volume; pptv= parts per trillion by volume

According to the consensus that is being established within the scientific community, these trace gases have to be associated with the gradual climate warming observed from the 19th century. It is believed that this warming is the result of disturbances in the energy balance of the Earth-atmosphere system directly caused by the increase of the atmospheric concentrations of greenhouse gases. The presence of aerosols[2] in the atmosphere tends to produce a cooling effect and to slightly reduce the rate of warming and even to cancel out the effects of the greenhouse gases at local scales. But, they cannot significantly thwart the effects of the greenhouse gases at a global scale since their lifespan in the atmosphere is very short.

Thus, there is an opinion, which explicitly attributes the warming trend to human influence because the increase of greenhouse gases mainly results from the use of fossil fuels, agriculture and land use changes. It is corroborated by many considerations and in particular by the outcomes of some attribution studies[3]. These have revealed some convincing correspondences between the observed patterns of atmospheric temperature change at different altitudes, seasons and geographical scales and the climate response to the combined effects of greenhouse gases and anthropogenic aerosols obtained in climate models (IPCC 1995).

However, there are clearly some difficulties and uncertainties to make accurate quantification of the amounts greenhouse gases released from the known sources or removed by the sinks. But it can be assumed that the atmospheric concentrations of the GHGs, especially carbon dioxide and methane, will keep on increasing to meet the needs of the economic activities. Furthermore, in spite of the official positions adopted during the international meetings, countries and people are manifestly not really inclined to change the direction of development and to adopt lifestyles that are less fossil fuel consuming. On the other hand, by considering the sources of the different GHGs, it can

---

[2] The aerosols correspond to small particles of dust, salt, soil and sulphates from fossil fuel combustion. About half the aerosols particles in the atmosphere are natural in origin, with the ocean as one of the most important sources. But, most of today's sulphate aerosols come from the burning of fossil fuels, destruction of tropical rain forests and other human activities (Whyte 1995)
[3] Attribution refers to the process of establishing cause and effect relations, including the testing of competing hypothesis (IPCC 1995)

be said that the industrialised countries have played the major role in the build-up of GHGs in today's atmosphere. The contribution from the developing countries is quite smaller. Deforestation in tropical areas represents an important share in this contribution. But most developing countries in particular those located in the tropical areas in which are the most important forest reserves do not like very much to discuss the issue of deforestation. They are particularly reluctant to hear criticism about the way the forests are managed and associate that question with their own sovereignty. In historical perspective, it is considered that the developing countries are responsible for less than 20% of the total industrial carbon dioxide emissions and for less than one third of the carbon dioxide and methane accumulated in the atmosphere (Oberthur and Ott 1999). Africa is the continent that has the lowest contribution of GHGs emissions. This one is estimated to be no more than 7% of the world total (Hulme et al 1995). However, it is believed that the developing countries in general and the African continent in particular will have a greater influence on the atmosphere during the current century because of land use practices and potential growth in energy consumption (Hulme et al 1995). The demographic variable and the economic activities will probably entail more pressure on the forests and an increased use of fossil fuels.

Figure 2.4: Carbon Dioxide Emissions of Industrialised and Developing Countries

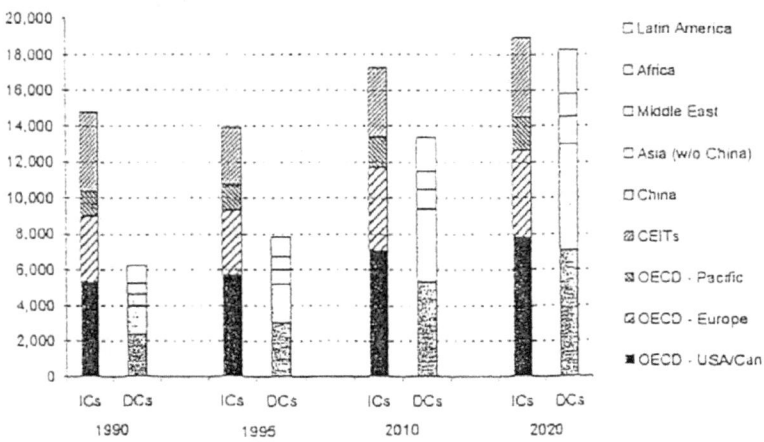

Source: IEA 1998 cited in Oberthur and Ott 1999

2.3 Doubts and Uncertainties

The abundant and deliberate use of terms such as "could", "would", "approximately", "estimates", "about" and other similar ones in the literature concerning the climate change issue expresses doubtless the humility and the wariness required in any scientific approach. On the other hand, it seems to reveal the amount of doubts and uncertainties still persistent in this field. As an emerging discipline, climate change impact assessment is currently developing in a context where the stirs and the opposite views are not rare. Such a situation is quite justified because in any circumstance, dealing with the future is not an easy exercise. Moreover, the complex nature of the climate system appears as a supplementary element that makes tricky the efforts to predict the exact effects of an increased atmospheric concentration of greenhouse gases.

Consequently, certain scientists express their scepticism and even consider with suspicion the projections made in the framework of the Intergovernmental Panel on Climate Change (IPCC). This institution is the body commissioned by the international community to assess the state of the scientific knowledge on the numerous dimensions of climate change. But in most cases, these climate sceptics are themselves exposed to criticism since they are supported by firms and companies such as Exxon and Mobil Oil known to be radically opposed to any efforts designed to change the current lifestyles and the direction of development. Because of this support, they are placed in an embarrassing situation that prevents them from claiming any credibility and they defend arguments or thesis that are favourable to the status quo (Oberthur and Ott 1999). For some other scientists, probably stirred up by an unbounded optimism, there is no need to get into a panic because the effects of climate change will be barely felt in the midst of other economic and technological changes. In other words, the development process and the future social, economic, political and technological changes will help to temperate or to overcome without pain the impacts of climate change (Rosenzweig and Hillel 1995).

Naturally, the no remedial policy that is suggested by the adepts of the wait-and-see attitude is unacceptable for the IPCC and for all the scientists sharing its vision of the future. For them, there is no doubt as for the existence of threats over the Earth's climate and subsequently, over human well being. Furthermore, the climate is even considered to have changed during the past century. That opinion is based on averred facts such as the global sea level rise by 10 and 25 cm, the global mean surface air temperature increase by about 0.3 to 0.6°C, the increase in rainfall over land in high latitudes of the Northern Hemisphere, etc. From the results of many studies, especially those comparing "the modelled climate response to combined forcing by greenhouse gases and anthropogenic sulphate aerosols with observed geographical, seasonal and vertical patterns of atmospheric temperature change", an human-induced effect on climate is becoming more and more evident (IPCC 1995). This changing climate under human influence is expected to continue in the future. However, in the IPCC's reports, the scientists carefully avoid adopting any peremptory tone even though they can legitimately claim that their positions reflect a scientific consensus. On the contrary, they recognise willingly that a lot of work has to be done and considerable scientific uncertainties have still to be reduced through further research. In that perspective, a comprehensive inventory of the different sources and sinks of greenhouse gases, the improvement of the understanding of the role of the oceans and their interactions with the atmosphere, a better modelling of the climate processes by the representation of cloudiness, etc... are issues that still need to be addressed by the researchers. The real challenges are to determine what could be the exact effect of anthropogenic greenhouse gases on the climate, to separate human influence and climate natural variability and to increase the accuracy or the quality of the predictions made concerning the regional impacts of climate change.

But one should not be naïve. Whatever is the amount of efforts and imagination invested in taking up these challenges, it will always be difficult suppress all the doubts and to obtain the unanimity.

Chapter Three
Climate Change on the International Scientific and Political
Agenda

3.1 Growing Concerns and First initiatives

The possible anthropogenic interference with the climate system is far from being a new idea just recently introduced into the debate to become one of the numerous global issues that mobilise and agitate the international community. It was evoked for the first time in the late nineteenth century by the Swedish scientist Svante Arrhenuis who "wrote a paper in 1896 in which he tried to correlate changes in surface temperature of the Earth with changes in atmospheric carbon dioxide. Seven years later, he noted that industry might put out enough carbon dioxide to actually warm the Earth" (Nilsson 1992). Thus, the first concerns about a potential climate change induced by human activity arose within the scientific community, which started long time ago to explore the real nature of the Earth's climate. Lectures, scientific workshops and conferences are naturally the principal opportunities used by the scholars, mainly climatologists, to exchange ideas, to express worries and to encourage further research and reflection. Progressively, an operational and well-organised scientific community has emerged at the international level and has showed a great ability to seize on the climate change issue, to "guess" its possible consequences and to communicate its worries. A converging set of observations has allowed the searchers to reveal a possible future climate change whose potential consequences might affect the whole planet. In order to get everyone ready for the necessary decisions to be made in the future, the international discussions on the climate which really started in the late 70s, were gradually widened enabling scientists and political leaders to work together (Roqueplo 1993). Such collaboration is still going on, especially because there is a strong need to underpin the political decisions by less scientific uncertainties. So, a series of events has stood out as milestones in the process of international mobilisation around the climate change issue. Mentioning some of them would probably help to improve the understanding of the historical foundations of the current debate on the perspectives of climate change.

- **1979**: First World Climate Conference also called "Conference of Experts on Climate and Mankind" co-organised in Geneva by World Meteorological Organisation (WMO), United Nations Environment Programme (UNEP) and the International Council of Scientific Unions (ICSU). The cooling of the Northern Hemisphere and the widespread drought and desertification in Sub-Sahara Africa are among the issues discussed by the participants.

- **1980**: The same organisations as above organised a meeting of experts in Villach (Austria) in order to assess the role of carbon dioxide on climate variations and to examine how the increase of greenhouse gases in the atmosphere could affect different regions of the Earth during the coming century. Besides, WMO and ICSU jointly initiate the World Climate Research Programme (WCRP).

- **1985**: A second scientific conference is organised under the aegis of the WCRP in Villach (Austria) in order to follow up and update the assessment of the role of carbon dioxide and other greenhouse gases in climate variations and associated impacts. The experts express fears concerning the occurrence in the first half of the coming century of a rise of global mean temperature greater than any other observed in man's history. In addition, a special recommendation is addressed to world leaders for policy actions.

- **1987**: The scientific assessment made on the role of carbon dioxide and other greenhouse gases in climate variations is used as a justification in two workshops organised in Villach (Austria) and Bellagio (Italy) to call for the development of international policies in response to possible climate change.

- **1988**:
  - Different governments are invited in Toronto (Canada) in order to participate in formal discussions that could lead to a possible "law of atmosphere" for the control of the atmospheric pollutants.

- The UN General Assembly adopts the resolution 43/53 on the "protection of global climate for present and future generations of mankind".

- A group of international experts called the *Intergovernmental Panel on Climate Change* (IPCC) is set up by the United Nations Environment Programme (UNEP) and the World Meteorological Organisation (WMO) with the mandate to assess the latest technical and scientific information related to the different aspects of the climate change issue.

Box 3.1: The Intergovernmental Panel on Climate Change (IPCC)

Without being structurally connected to the United Nations Framework Convention on Climate Change, the IPCC is the main institution providing the scientific impetus necessary to the perpetual consolidation of the international regime on climate change. Bringing together leading scientists from all over the world to assess the scientific, technical and socio-economic information relevant for the understanding of human induced climate change, its potential impacts and options for mitigation and adaptation, the IPCC is structured around one task force on "Greenhouse Gases Inventories" and three thematic Working Groups (I, II, III) on "the science of climate change", "Impacts, adaptation and vulnerability" and "Mitigation of climate change". Considered as the most credible source of information on climate change, it has already produced two assessment reports in 1990 and 1995. The former contributed to launch the negotiations on the UNFCCC while the latter influenced a lot the process having led to the adoption of the Kyoto Protocol. Its third assessment report is expected in 2001. In addition, the IPCC publishes "technical papers" and "Special reports" for one of the two standing bodies of the Convention, namely the Subsidiary Body for Scientific and Technological Advice (SBSTA), which is its link to the UNFCCC.

Source: Adapted from http://www.ipcc.ch

- **1990**:
  - The IPCC publishes its first report in which it defines Climate Change as a global threat and calls for a global treaty to deal with the problem.
  - The second World Climate Conference is organised in Geneva (Switzerland) and the participants subscribe to the views expressed in the first IPCC's report.
  - The UN General Assembly set up an Intergovernmental Negotiating Committee (INC) and assigns it the mandate to conduct the negotiations for the elaboration of a Framework Convention on Climate Change with appropriate commitments.

- **1991/1992**: From February 1991 to May 1992, five meetings are organised in Virginia, Geneva (twice), Nairobi and New York by the Intergovernmental Negotiating Committee, which finally reached the conclusion that "a flexible, voluntary response by Nations to reduce net concentrations of greenhouse gases would be the backbone of the Climate Convention" (Morrissey 1998).

3.2 Development of an international regime on Climate Change

In March 1989, different heads of State and Government gathered in The Hague (Netherlands) to express their common concern about the existence of dangers for the natural ecosystems and the interests of mankind due to the warming of the atmosphere and the depletion of the ozone layer. They strongly stated that the search of solution was vital, urgent and global. In the declaration they signed at the end of the meeting, which is commonly known as the "Declaration of The Hague", there was an implicit recognition that the existing international legal system was inadequate to address the climate change issue. Consequently, the international community was urged to develop "new principles of international law including new and more effective decision making and enforcement mechanisms". In the view of the signatories of the declaration, new legal instruments such as a framework convention and a new institutional authority within the framework of the United Nations were necessary to take into charge the combat against further global warming and climate change. So, the decision made one year later by the General assembly of the United Nations to establish the negotiating committee mentioned above can be appreciated as a positive response to the pressing call contained in The Hague declaration. Started in February 1991, the negotiations for the elaboration of a Framework Convention on Climate Change ended fifteen months later by a compromise between various positions or interests represented by different groups of countries and the adoption of the new international regime.

3.2.1 The United Nations Framework Convention on Climate Change

Once adopted by the Intergovernmental Negotiating Committee (INC) on May 1992, the United Nations Framework Convention on Climate Change (UNFCCC) was opened for signature the month after at the World Conference on Environment and Development in

Rio de Janeiro (Brazil). The ratification process by countries developed rapidly and on May 1994, the convention came into force. At the present time, the European Union and more than 180 countries are parties to the convention. This rapid success is probably due to the fact that the convention is conceived to tackle a common concern of humankind and also because most countries were willing to endorse the objective and the principles defined in the convention.

Box 3.2: Objective and principles of the UNFCC

The ultimate objective targeted by the convention and defined in its article 2 is to stabilise atmospheric concentrations of greenhouse gases at a level that would prevent dangerous anthropogenic interference with the climate system. Though not quantified, that level should be achieved within a time frame sufficient to allow ecosystems to adapt naturally to climate change, to ensure that food production is not threatened and to enable economic development to proceed in a sustainable manner. All parties are committed to contribute to the achievement of this objective individually or co-operatively by promoting policies and measures on the basis of the principles of equity and precaution and in a perspective of sustainable development.

Source: adapted from http://www.unfccc.de/index.html

Consequently and in accordance with "their respective capabilities and their common but differentiated responsibilities" with regard to the occurrence of climate change, the country Parties are divided into two groups:

- The Annex I Parties comprising all the industrialised countries (the European Union, the OECD members in 1992 and the Countries with Economies in Transition (CEITs) such as the Russian Federation and other Eastern European countries) which have historically contributed the most to global warming. For this reason, they are committed to take the lead in addressing climate change. Thus, they are subjected to the *not legally binding obligation* to reduce their greenhouse gases emissions to 1990 (or another year for the CEITs) levels by the year 2000. In addition, they have to submit regular reports detailing the actions undertaken to achieve the common objective. The OECD members in 1992 are also listed in a second specific group of Parties called Annex II Parties which are subjected to the special obligation to facilitate the transfer of climate-friendly technologies to both developing countries and CEITs. Besides, they have to help specifically the developing countries with additional financial resources to tackle the climate change issue.

- The non-Annex I Parties comprising the developing countries have relatively negligible contribution to the warming effect. Hence, they are just required to submit national communications in which their emissions inventories and the actions achieved to combat climate change are reported. To do it, they must receive a financial assistance from the convention.

As a legal instrument conceived to develop constantly "in the light of the best available scientific information and assessment on climate change and its impact, as well as relevant technical, social and economic information" (article 4.2d), the convention is regularly reviewed by its most important institution: the Conference of Parties (COP). Acting as the convention's decision making body, the COP is keeping under its authority and guidance two additional bodies: the Subsidiary Body for Scientific and Technical Advice (SBSTA), which often plays the role of interface between the COP and the IPCC, and the Subsidiary Body for Implementation (SBI).

Technology transfer and financial assistance to developing countries are declared to be two critical factors for the successful implementation of the convention. With this regard, the convention's financial mechanism is being operated by the Global Environment Fund (GEF) and the developing countries require more fairness, equity and transparency in the functioning of this institution.

For many countries and especially for those particularly vulnerable to the adverse impacts of climate change, the soft nature of the obligations assigned to the Annex I Parties and the subsequent lack of provisions for non-compliance procedure cannot allow a successful implementation of the convention. The first national communications submitted by the industrialised countries raised doubts and concerns as to the adequacy of their commitments with regard to the objective to reduce greenhouse gases emissions at 1990 levels by the year 2000. The review of the actions undertaken in these countries revealed their insufficiency. Consequently, an increasing number of non-Annex I Parties started militating in favour of a strengthening of industrialised countries' commitments. The Alliance of Small Island States (AOSIS), one of the most resolute groups for the assignment of substantive commitments to industrialised countries, took the lead of this new diplomatic battle. In a context marked by some resistance from countries like the United States of America and the Organisation of Petroleum Exporting Countries (OPEC) against the strengthening of the industrialised countries' obligations, The IPCC Second Assessment Report came out with the following disconcerting argument: "the balance of evidence suggests that there is a discernible human influence on global climate" (IPCC 1996). Naturally, this statement contributed to reinforce the determination of those pleading for the application of legally binding obligations to Annex I Parties. Despite of the diversionary actions desperately undertaken by the OPEC, the first session of the Conference of Parties held in Berlin in 1995 provided the opportunity to engage the international regime on climate change in a new process, which led to the adoption of the Kyoto Protocol.

3.2.2 The Kyoto Protocol

The so-called Kyoto Protocol is an additional legal instrument to the United Nations Framework Convention on Climate Change (UNFCCC). Adopted at the third session of the Conference of Parties on December 1997 in Japan, it is the result of two and half years of hard negotiations between countries willing to further elaborate the international regime on climate change and countries reluctant to accept new commitments. Concretely, it commits the convention's Annex I Parties listed in the Protocol's Annex B[4] to differentiated legally binding obligations for the reduction or limitation of the emissions of six main greenhouse gases: carbon dioxide, methane, nitrous oxide, sulphur hexafluoride, hydrofluorocarbons and perfluorocarbons. The objective is to reduce the industrialised countries' aggregate greenhouse gases emissions by at least 5% below 1990

---

[4] Annex B Parties include the following: Australia, Austria, Belgium, Bulgaria, Canada, Croatia, Czech Republic, Denmark, Estonia, European Community, Finland, France, Germany, Greece, Hungary, Iceland, Ireland, Italy, Japan, Latvia, Liechtenstein, Lithuania, Luxembourg, Monaco, Netherlands, New Zealand, Norway, Poland, Portugal, Romania, Russian Federation, Slovakia, Slovenia, Spain, Sweden, Switzerland, Ukraine, United Kingdom of Great Britain and Northern Ireland, United States of America.

levels in the commitment period 2008-2012. This is maybe an interesting achievement especially if one considers the lack of certainty that surrounds the climate change issue and the resistance developed by the numerous sceptics. But the Kyoto Protocol as the whole international regime on Climate Change is still considered as an "unfinished business" (Oberthur and Ott 1999). This feeling derives from the fact that the Kyoto Protocol has taken up again or established innovative tools that need to be further elaborated and are still under discussion. These are the flexibility mechanisms known as Emissions Trading (ET), Joint Implementation (JI) and the Clean Development Mechanism (CDM). They are all under design in order to help the countries subjected to legally binding obligations to reduce the costs of achieving their emission targets by implementing actions in other countries where they are cheapest. However, each of them possesses some specificity that distinguishes it from the two others. Without entering into the details, they can be briefly defined as follows:

- Joint Implementation mechanism is originally introduced into the climate change regime in 1993 on the basis of the FCCC. At the first Conference of Parties in 1995, a pilot phase of Activities Implemented Jointly (AIJ) involving industrialised and developing countries was established. The concept is taken up again by the Kyoto Protocol and its use is to be limited to industrialised countries only. It allows an Annex I Party to transfer to, or acquire from, another Annex I Party emission reduction units obtained through the execution of projects designed to reduce greenhouse gases emissions or to enhance their removal by sinks. The private sector is expected to be the main operator of JI projects and to catch up the opportunities provided for cheap emission reductions. Discussions are still underway in order to define, among other issues, the model of JI. While certain countries prefer to privilege a bilateral concept in which the participating Parties would be responsible for the verification of the whole process, others seem to support a multilateral option with the involvement of the institutions belonging to the Climate Change regime.

- Emissions Trading is a mechanism allowing the Parties listed in the Protocol's Annex B to participate in a trading system in which a country can sell part of its assigned amount or the credits acquired through JI projects, to another country, which has exceeded its allowed emissions levels. So, emission allowances and emission reduction units are the "commodities" that will be used for such a trade. Although there is no decision concerning the participation of the private sector, that possibility which already exists in the JI mechanism raises some apprehension because of the risk of having a too large trading system where it would be difficult to monitor and to enforce the rules.

- The Clean Development Mechanism (CDM) is the tool of interest for the developing countries, especially the African countries. Contrarily to the precedent mechanisms where only industrialised countries are involved, this one is in principle intended to favour the involvement of the developing countries in Kyoto Protocol. Defined in article 12 of the Protocol, the goal of the CDM concerns the developing countries as wells as the industrialised nations. It is question to help developing countries to follow sustainable development paths through the promotion of environmentally friendly project activities funded by industrialised country government or business community. By backing such projects, the Annex I Parties would acquire certified emission reductions (CERs), which are considered to correspond to part of their obligations.

These tools, Joint Implementation, Emissions Trading and Clean Development Mechanism are still under discussion, especially for setting up the governance system and the detailed rules. The coming Conference of Parties of the UNFCCC, the sixth, normally scheduled to take place in The Hague (Netherlands) on November 2000 is likely to offer the opportunity to adopt the operational details concerning the flexibility mechanisms. But, it is already admitted that they cannot be used by any Annex I Party as an exclusive contribution to the implementation of the Kyoto Protocol. On the contrary, they are to be used in addition to the package of domestic measures carried out in order to reduce the emissions of the six GHGs covered by the Kyoto Protocol. Since negotiations like those engaged for the final design of the CDM are always hard because of the determination of groups of countries to impose their views and interests, it is not easy to guess the real possibilities that it will provide the developing countries. But it is essential for these countries and especially those in Sub-Saharan Africa to seize the opportunity offered by article 12 of the Protocol, which stipulates clearly that "the purpose of the CDM shall be to assist Parties not included in Annex I in achieving sustainable development and in contributing to the ultimate objective of the convention (…)". So, climate change is implicitly regarded as a constraint to sustainable development. Consequently, the strategy for Sub-Saharan Africa within the international regime on the climate should be to set as its major objective the promotion of sustainable development and to use all the opportunities that are made available by the UNFCCC and the Kyoto Protocol to achieve it.

Chapter Four
Virtual Implications and Potential Repercussions of Climate Change at Global Scale

As an irreplaceable source of goods and services, the natural environment supports clearly life on Earth. It is of critical importance to humans and to all other species. But it does not exist by itself since it is shaped and sustained by the Earth's climate patterns, which determine its physical characteristics and its biological wealth. That is why the planet is characterised by a diversity of ecosystems belonging to different climatic zones and human society is enabled to promote various types of activity for its survival and its well being. In these conditions, any change in the climate patterns under the influence of natural or human factors would generate effects on the environment and would affect in one way or another people's social and economic status. Such a situation is very likely to happen because evidence is mounting that mankind is disrupting the global climate. The global temperature that largely governs the earth's climate patterns has been gradually rising for more than one century and this seems to be linked to the growing atmospheric concentrations of greenhouse gases. As a consequence, global warming might alter the climate patterns, which could, in turn, affect the nature in many parts of the world and generate or increase human suffering and disarray (UNEP 1993).

However, the scientists seem to be unable to provide detailed indications of what will occur at different geographical scales. They cannot give precise indications concerning the extent and the timing of the events. The computer models used to make the predictions cannot catch and reproduce all the processes that govern the climate system. Hence, the scientists' confidence in their ability to predict the potential impacts of climate change is not unshakeable (Nilsson 1992). Because of uncertainties over the details, they cautiously prefer to draw a broad picture of the major impacts that might result from climate change. With this regard, an outstanding mobilisation exists within the IPCC and efforts are still being made in order to identify the vulnerable ecosystems, resources and activities and the extent to which they might be affected. In Section 4.1, the parts of the nature and the dimensions of human life that are in danger because of climate change will be examined. But on the basis of the links existing between climate, ecosystem and man, it is already clear that the perspective of climate change places under threat the natural environment as wells as the society.

4.1 Threats over the environment

As climate change is expected to result from the fact that the Earth is gradually getting warmer, a range of effects is likely to occur under the direct or indirect influence of rising temperatures. In the whole literature dealing with the climate change issue, sea level rise is seen as an inevitable outcome of global warming. In response to the thermal expansion of the ocean waters and the melting of the land-based ice mass, mainly located in Greenland and Antarctic and the mountain glaciers, the level of the sea throughout the world is expected to keep on rising. Consequently, a range of interrelated impacts including erosion, inundation, increased flooding and salinisation of soil and freshwater might affect the coastal lowlands and low-lying islands. These areas and more generally the coastal zones are known to shade a rich diversity of ecosystems and to be propitious to the development of socio-economic activities. That is why probably about half of the world population lives in such places (IPCC 1998). But in a context of climate change, they might be less attractive and even become dangerous for the reasons mentioned above. So, the coastal wetlands, the mangrove, the species and ecosystems supported by the permafrost, etc might pay a heavy tribute to climate change.

However, such disasters would never be a novelty for the world. Coastal flooding, inundation and erosion are current phenomenon throughout the world because of the natural vulnerability of certain countries. Every year, tens of millions of people are at risk. Bangladesh is in this regard a martyred country for the high incidence of flooding and the extent of the damages. Climate change will probably exacerbate the problems in areas prone to flooding. Consequently, the ecosystems located in the coastal zones and their biological wealth would be lost or considerably affected by sea level rise.

Concerning the fauna and the flora, changes have to be expected in the composition of species within both terrestrial and aquatic ecosystems. The combination of altered rainfall patterns, increased evapotranspiration and reduction of soil moisture could shift vegetation zones and alter productivity and species composition in rangelands and in forested systems especially under the tropics. Naturally, the wildlife that is supported in such zones should be affected by the disruption of their habitats. The species that would not be able to adapt to new ecological conditions or to find another suitable environment might be condemned to disappear. So, it is reasonable to believe that global biodiversity will be affected in one way or another by climate change (UNEP 1993).

In the oceans, climate change could lead to disturbances affecting the dynamic of the currents and subsequently the distribution of nutrients, the biological productivity and the role of the oceans as carbon sink. The regional differences in sea-surface temperature and the temperature difference between air over oceans and that over the continents could respectively alter the trade winds or generate stronger along-shore winds.

Another topic of big concern is the potential impacts of climate change on the water resources. In several parts of the world and particularly in arid and semi-arid areas, water is already considered as a scarce element and a source of conflicts between communities. The combination of drought[5] and demographic growth is currently reducing the per capita quantities available and exacerbating the problem. In a context of climate change, especially when it is accompanied by a decrease in rainfall, the situation might become worst.

By reducing the snow and ice cover and the supply of rivers and groundwater, global warming could increase the scarcity of water at a global level. Chronic shortfalls of water would certainly aggravate drought, which is a recurrent problem in the arid and semi-arid areas and probably deserts would expand. Other areas might escape drought but they could face disasters such as flooding (IPCC 1998). In the recent years, the increase of

---

[5] Drought is generally seen to be an abnormal reduction in water availability; moisture supplied below average for short periods of 1-2 years (Agnew 1995).

rainfall has led many European rivers to run out of their beds causing inundation of cities and rural landscapes. Maybe this is a signal of what might be in the future more common events.

4.2 Threats over human society

In the light of the considerations presented above concerning the potential impacts of climate change on the environment, human society's vulnerability to global warming appears as an indisputable contingency. When the land, the oceans, the water resources, the biological diversity and the health status of people, which constitute the basis of human activity are placed under the threat to undergo damages, there is then less hope for progress, happiness and the perspectives of sustainable development. The heavy burden that climate change could impose on certain societies through hunger, natural disasters, high morbidity and mortality from infectious diseases such as malaria and cholera would in areas throughout the world, especially under the tropics, add to the fragility of situations already precarious. The vision of the potential impacts associated with climate change could be entitled 'chronicle of an announced pain' if there were not the fear to be extravagant and exaggeratedly alarmist. As what is likely to happen at the expense of the environment, the burden on society will be heavy and unevenly distributed. Concretely, in the same way that species and ecosystems might be affected by water scarcity, drought, floods and others, humans would also have to pay a tribute to these phenomena through disturbances or destruction affecting water supply, energy supply, agriculture, fisheries and food supply, tourism, settlements and the economic infrastructures, etc. With this regard, the risk of hunger is considered to be high in certain regions though the global agricultural production would be sufficient to feed the world population (IPCC 1995). The possible shortage of food in the tropics where rain-fed crops and subsistence farmers are the main features of the agricultural system may be compensated by a possible increase in productivity in other areas, probably in middle and high latitudes. Fisheries will possibly be affected in the same way as agriculture. In spite of the vulnerability of the sector to climate change and the overexploitation of the resources in certain seas, the world production is likely to be maintained at levels consistent with the global needs. But some localised places might loose their status of production centres.

Considering that 800 million people are currently exposed to malnutrition and 1.3 billion people do not have adequate access to safe water, one would easily realise that the perspectives in which climate change is likely to engage agriculture and fisheries should raise a lot of concerns among the people and the countries at risk (IPCC 1998). Through agriculture, fisheries and other new flourishing activities such as tourism or strategic sectors like energy supply e.g. hydropower, it is feared that climate change would have detrimental effects on the economic activity in many regions, notably in the poor countries. Their economic potential might be subjected to damages that would seriously compromise development take off. As a consequence, entire populations might be obliged to leave their homes and to increase the streams of migrants, refugees or displaced people.

Besides, in addition to aggravating the disparities in the world with regard to adequate access to food and potable water, climate change can also favour the resurgence and the spread of diseases beyond the boundaries in which they are currently confined. Most of the environmental impacts of climate change such as flood, drought, wind storms, etc described above may directly affect human health through injury or violent death or indirectly by weakening people to an extent that they cannot oppose an appropriate resistance to diseases. They may also set new ecological conditions in which the development of pathogenic seeds would be made easier. The areas considered by most studies to be at high risk are the tropical and subtropical regions. Malaria and cholera are some of the diseases that raise the biggest concerns since they might in response to environmental changes, enlarge their zones of incidence. Malaria is a known redoubtable disease that is far from being under control in spite of the efforts made by the international community through the World Health Organisation (WHO). It is the prime cause of morbidity in many developing countries. Approximately, 350 million new cases appear every year and 5% of the world population are currently infected (WHO 1996). These figures are absolutely dismaying and it would be natural to get worried because of their possible increase under the influence of climate change.

Chapter Five

Impacts on Sub-Saharan Africa: New Challenges and Exacerbation of Recurrent Problems

5.1 Vulnerability of Sub-Saharan Africa

Because of widespread poverty, Sub-Saharan Africa can certainly be regarded as the most vulnerable region to climate change (IPCC 1998). This is not surprising. It is founded on the historical and current situation of Africa, which points out the continent as the greatest challenge to world development. The people and the environment on which they depend seem to be incurably associated with problems. This raises the feeling that tragedy is a label of the African way of life. Practically, all aspects of human life that are likely to be affected by climate change are already placed under red indicators in Africa. For this reason, people can be inclined to believe that climate change or its effects have already started to affect Africa. A multidimensional crisis is bruising the continent for decades and is leading to a steady decline of the living standards while the rest of the world seems to be moving forward.

Environmental bankruptcy can legitimately be considered as the central element and the driving force of the African tragedy and its social and economic nightmare. This is particularly true for the ordinary people who depend for their survival only on the resources of the land. Because of climate variability, demographic pressure and probably bad policies and practices, the African environment is constantly ill-treated and that reinforces the breakdown in the relationship between people and life support systems. Soils and vegetation are being impoverished, food production is gradually declining, poverty is gaining ground, and diseases are spreading (Timberlake 1985).

Table 5.1: Number of natural disasters in the period 1970-1994 by global region

Region	Africa	America	Asia	Europe	Oceania
Drought and Famine	272	51	88	15	15
Flood	168	373	628	138	139
High Wind	84	428	683	214	200
Landslide	12	87	96	21	10

Source: WHO 1996

5.1.1 Drought, Water Shortage and Desertification

Those who pay attention to the history of Sub-Saharan Africa know that drought is one its most important scourges. Recurring at unprecditable intervals, it is at the origin of ecological and human disasters that periodically martyr entire regions of the continent. Recently in the early 90s and particularly in 1992, a severe drought affected most countries of Southern Africa. The cereal production dropped to less than 50% of the requirements and about one quarter of the population of the region was in need of food assistance. That drought was believed to be associated with the El Niño Southern Oscillation (ENSO) phenomenon (Myers and Kent 1995). This originally refers to a tepid current that arrives every Christmas to the coasts of Peru and Ecuador and gives some seasonal time off to the local fishermen by driving away fish. It has become now warmer and long-lived and has a big impact not just locally on the coasts of Peru and Ecuador but around the world. It is considered as a world weather catastrophe because it brings droughts and crop failures to some parts of the world and flooding to others. Southern Africa seems to be particularly exposed and vulnerable to the effects of the ENSO phenomenon (Xoagub 1997).

Figure 5.1: Areal Extent of the 1991/1992 Drought in Southern Africa

Source: Zinyowera and Unganai 1992 cited in Hulme et al 1995

However the region that can be really pointed out as the symbol of the drama associated with drought is the southern boundaries of the Sahara Desert better known under the name of Sahel. Its rainfall is often meagre and follows erratic trends. During the last decades, it has been regularly struck down by repetitive droughts producing effects that are still lying heavy on the ecosystems as well as on people's livelihoods (Agnew 1995). From the 60s, annual rainfalls started decreasing at levels 20 to 40% lower than those of the period 1931-60. In particular years such as 1984 and 1990, the decrease was more important, falling below 50% of the rainfall obtained during the colonial period. For this reason, the term desiccation[6] has come into widespread use to describe the situation prevailing in the Sahelian region. Different hypotheses have been developed to explain that process of aridification. Some specialists consider that land cover changes through deforestation and desertification and sea surface temperatures in the Atlantic Ocean have

---

[6] Desiccation (or aridity) refers to a process of aridization lasting decades. It involves a long-term decline in moisture supplies (Agnew 1995).

to be associated with the decrease of rainfall in the Sahel and even in the whole tropical north Africa (Hulme et al 1995).

Figure 5.2: Frequencies of severe droughts in the Sahel and Sudan, 1901-1989

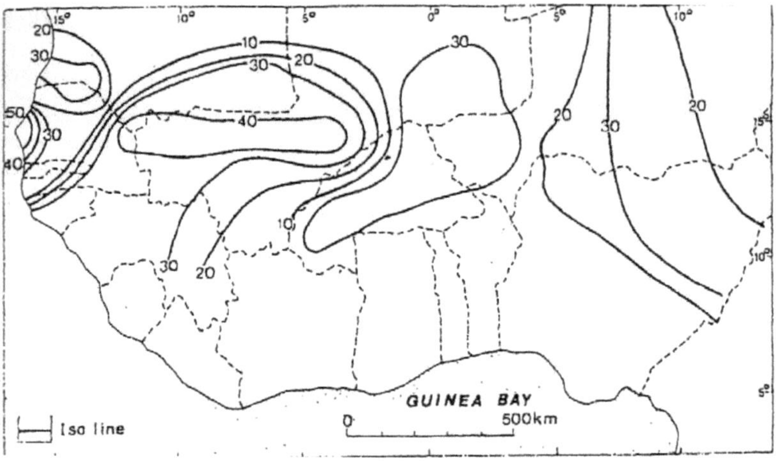

GUINEA BAY
500km

Iso line

Source: Ojo 1997

Naturally, the main characteristic of drought is that it leads to scarcity of water resources. This generates situations of water deficits, which are detrimental to humans as well as to animals and plants. It is believed that water will be one of the most pressing resource issues during the current century. Presently, several countries throughout the world are subjected to chronic problems of water shortage. Not surprisingly, most of these countries are located in Sub-Saharan Africa because of an exceptionally variable precipitation across much of the subcontinent and a very low runoff. However the situation is heterogeneous within the region. Some countries, especially those in central Africa have enough water but for most countries, demographic growth and economic development are creating an imbalance in demand and supply. This can lead to conflicts within and between countries. In 1997, the International Green Cross expressed its big concern about the tension prevailing in many African river basins such as the Nile, Senegal and Volta (Spore CTA 1997). About 300 million Africans are supposed to be living in a water-scarce environment and the perspectives for the coming decades are quite alarming because the population is expected to exceed one billion people by 2025. Moreover, the outstanding potential for hydropower generation estimated to be about 1.4 million GWh per year is currently under exploited. In 1989, the total hydro-electricity production was only 41,000 GWH (World Bank 1996). This potential is naturally vulnerable and might be considerably lessened by the frequent cycles of drought. Generally when the per capita water availability is less than 1000m^3, water scarcity is then considered to be a serious constraint for socio-economic development and for the protection of the environment (FAO 1997). That is already the situation in many African countries and in case of drought, they are a prey to additional social and economic problems.

Figure 5.3: Per Capita Water Availability in some African Countries

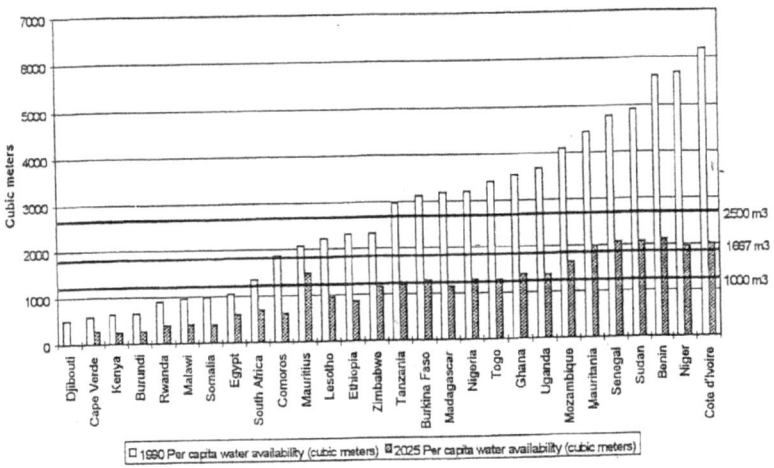

Source: World Bank 1996

38

Box 5.1: Drought and Desertification in the Sahel Region

Located in Sub-Saharan Africa, the Sahel extends over the meridional fringe of the Sahara Desert between 13° and 17° of latitude north. The climate is characterised by two seasons: a long dry one between October and June and a short rainy season lasting from July to September with precipitation that averages between 250 and 500 mm. During the last decades the region was afflicted several times by severe droughts. That of the period 1969/73, killed 100,000 persons in Ethiopia, brought about a drop of 2 to 15 meters in ground water table levels, decreased the flow rates in all rivers, 25% of the cattle died or were slaughtered, hundreds thousands of people became destitute and dependent of food aid. In the 80s, about 10 million people were forced off their land and obliged by the circumstances to settle in refugee camps or to migrate towards the urban areas. Associated with overcultivation and overgrazing, drought is fuelling the desertification process that is gradually reducing the productivity of the land. Since 1925, more than 650,000 km² of land have been turned into desert along the southern fringe of the Sahara.

Source: Myers and Kent 1995; World Health Organisation 1996; World Bank 1996

If the correlation between land cover changes and rainfall patterns is a correct hypothesis, that means the worst is still to come for Africa because of the dynamics of deforestation and desertification in the continent. Various factors including drought, demographic growth, agricultural expansion and increasing demand of timber and woodfuel are converging to put the African forests under a strong pressure. It is estimated that annually 3.7 million hectares of forests and woodlands are erased from the surface of the continent and 20 to 50 million tonnes of soil are carried away by wind and water erosion. Among the 900 million people throughout the world who suffer from the effects of drought and desertification, 200 million are in the African countryside, particularly in Sub-Saharan Africa, which is the most affected area (Mainguet 1999). With such conditions, it is not surprising that malnutrition, famine and poverty are deep-seated in the African continent. The climate instability with frequent outbreaks of drought combined with demographic pressure and a high dependence to rain-fed agriculture places the African societies and economies in a constantly precarious situation.

5.1.2 Poverty, Malnutrition, Diseases and Public Health

With 12% of the world population, the African continent seems to be a land of predilection for poverty. Among the 36 countries considered to be the poorest in the world, 29 are located in Africa, essentially in the Sub-Saharan region. Whatever the form of poverty that one takes into consideration, income poverty or the broad index of human poverty, in which several forms of deprivation such as low life expectancy and illiteracy are included, Sub-Saharan Africa remains the subcontinent where the situation is more worrying (WRI et al 1998). For decades, the per capita Gross Domestic Product (GDP) is stuck under $500. Although the world economy has grown approximately fivefold since the end of the Second World War, most countries in Sub-Saharan Africa have rather experienced economic decline or stagnation. Among other factors that could be used to explain this economic decline or stagnation in the subcontinent, are the high dependence on exports of primary commodities, the instability of the commodity prices on the world markets and the burden of the foreign debt. For Sub-Saharan Africa, the volume of the exports that comprise 90% of commodities increased by 25% in the 80s, but paradoxically the revenues decreased by 30% (Myers and Kent 1995). At the same time, the payment of the debt service, estimated at $10 billion per year, prevents the region from undertaking significant investments in strategic sectors such as Health and Education.

Figure 5.4: Poverty in Sub-Saharan Africa 1985-2000

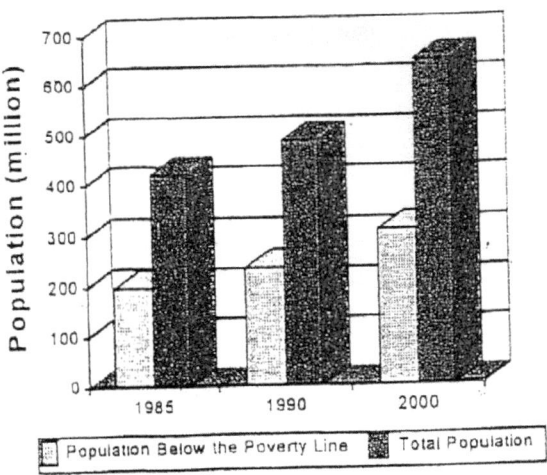

Source: World Bank 1996

40

However, environmental bankruptcy and uncontrolled demographic growth are believed to be among the most important determinants of the African distress. Most of the African poor people live in the rural areas. To meet survival needs, they rely mainly on the land (soil, water and vegetation), which constitutes their economic capital. However as is already well known, the land in Africa is one of the most subjected to degradation[7] and which makes it more and more unable to sustain the population. Even though food production has grown during the past decades at a rate of 2% per year, the farmers who practise mainly rain-fed agriculture are absolutely unable to keep pace with the population growth of 3% (World Bank 1996). To compensate for the decline of the per capita food production, they are destroying the resource base through overexploitation of the soil and agricultural expansion into forests, thereby compromising the means for future survival. But Sub-Saharan Africa is currently experiencing a food deficit of more than ten million tonnes per year. Because of widespread poverty and the lack of money, it is naturally difficult for people to buy food and to have access to an adequate and varied diet to ensure healthy living. Hence, Sub-Saharan Africa is a propitious land for undernutrition, which is continuously spreading. About one quarter of the 840 million people suffering from chronic undernutrition in the developing world, lives in Sub-Saharan Africa. Moreover, the projections made by the United Nations Food and Agriculture Organisation concerning the region are not disquieting at all because there is no improvement at the horizon 2010 (WRI et al 1998).

---

[7] Land degradation is a reduction of the land to support a particular use (Blaikie and Brookfield 1987 cited in IPCC 1998). It is also defined by Agnew (1995) as a persistent decrease in the productivity of vegetation and soils.

Figure 5.5: Trends in undernutrition in Developing Countries 1969-2010

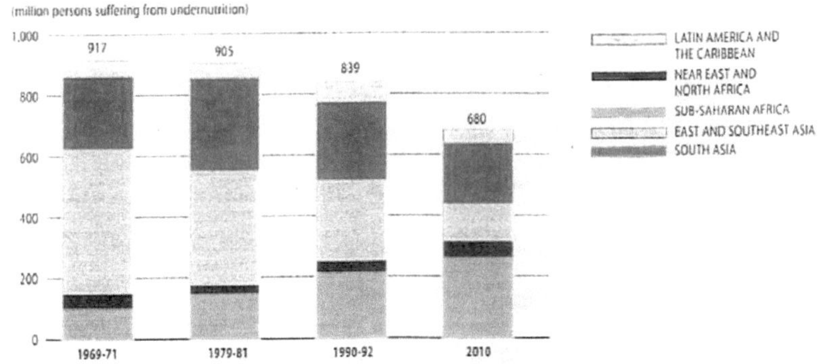

(million persons suffering from undernutrition)

Source: World Resources Institute, 1998

5.1.2.1 Inadequate Government Policies

Generally, people accept easily the idea that in Africa, food deficits and related problems result only from environmental degradation. This is not always true. At times, government policies should also be stigmatised because in 1983-84, the Sahelian farmers obtained a record harvesting of 154 million tonnes of cotton fibre (compared with 22.7 million tonnes in 1961-1962) whereas they were unable to produce enough grain to feed themselves (Timberlake, 1985).

5.1.2.2 High Incidence of Diseases

Another critical problem in Sub-Saharan Africa that is interrelated with poverty, undernutrition and environmental stress is the high incidence of diseases, particularly the infectious diseases. These are principally water-related. The unavailability of water in sufficient quantity and quality and the poor sanitation services are the main determinants of the worsening health status of African people. They create propitious conditions for the spread of diseases such as diarrhoea, cholera, malaria, etc. Generally, the young generations, the children in particular, are the first victims of these maladies. In 1996, Diarrhoeal diseases killed 2.5 million people throughout the world and Africa was the continent with the highest number of victims; it is currently having the highest diarrhoea death rate with 17/1000. About 80 million African people would be at risk for being affected by cholera and the refugees or displaced persons are the category of populations who use to pay the heaviest tribute to that disease. Malaria is also claiming the lives of people in similar proportions. Approximately, one to three million people are victims of this disease mainly in Sub-Saharan Africa every year. The region possesses the conditions that favour the spread of the disease; heat, poverty, inefficient waste disposal, etc. In addition to the number of victims killed by malaria, there is also the morbidity rate and the cost it represents. Rightfully called the laziness disease because of its debilitating effect, malaria annually affects more than 300 million people world-wide and deprives the meagre African economies of the contribution of workers hit by the illness, especially during the rainy season (WRI et al 1998).

Figure 5.6: Malaria distribution in Africa

African cities
Severe drug resistance,
increasing deaths in
young adults.
Inadequate sanitation,
overburdened services.

Main areas where malaria transmission occurs
(circled areas represent islands where detail cannot be shown)

Source: WHO, 1996

Whatever the impacts of climate change on Sub-Saharan Africa may be in the coming decades, they would surely occur in a context already hard for humans, the environment and economic activities. As has already been mentioned, the continent is entangled in

43

inextricable problems. The data in this chapter reveal that the situation prevailing in Africa has in some regards the features of a real disaster. Many countries are unable to find the necessary resources to fight any additional environmental catastrophe. They are in a state of ecological and economic fragility that makes them more than ever vulnerable to any kind of disturbance. The frequent outbreaks of drought and diseases did not leave the continent any kind of capability to deal with such events. The countries are certainly unprepared to deal with climate change.

5.2 Expected Impacts of Climate Change on Africa

In the whole literature concerning the potential impacts of climate change especially at the regional or local scale, the tone employed by the scientists involved in this area of research is generally stamped with a lot of caution. The researchers are conscious that climate change is not an easy topic; it is therefore unwise to claim the slightest certainty on this subject. It is absolutely impossible to accept that the computerised models used by the scientists to reproduce the various processes involved in the climate system and to construct on the basis of scenarios their predictions about the potential impact of the greenhouse effect, lead to irrefutable findings. At best, the results obtained from these complex models represent a broad indication of some possible impacts whose probability and magnitude cannot yet be certified. While waiting for the improvement and the refinement of the models for more accurate predictions, the scientists generally decline to go beyond qualitative discussions of the potential impacts of climate change (Hulme et al. 1995). Without being necessarily sufficient, the results of a qualitative assessment of the impacts of future climate change may be considered as an acceptable basis and a good source of inspiration for the elaboration of national and international responses to global warming.

In this respect, the work carried out within the framework of the IPCC has largely contributed to encourage the international community to go ahead in elaborating a regime on climate change and the industrialised countries to accept the principle of evolving towards further commitment. There is less and less contesting about the scientific validity of the IPCC's analysis and reports. These reflect a large consensus and represent the most important share in the leading literature on the climate change issue. The content of this chapter on the expected impacts of climate change on Sub-Saharan Africa is mainly inspired by the IPCC's work. Relying on other works may be hazardous since the climate models used do not necessarily lead to convergent results. In the context of the IPCC, the estimates got are obtained through simulations. They derive from changes in the climate system likely to occur in case of an equivalent doubling of the atmospheric concentration of carbon dioxide. In such a case, both global mean temperature and mean sea level are expected to follow a rising trend respectively in the range 1-3.5°C and 15-95 cm by the year 2100.

It is believed that these changes will have diverse consequential effects on the current situation in Africa. As was explained previously, the African ecosystems that should normally serve as springboard for the promotion of a sustainable socioecomic development are already subjected to a continuous degradation. Despite the difficulties currently encountered by the scientists in their attempts to know how exactly the climatic system in the different regions of the world will respond to global warming, it seems that the influence of climate change will worsen the situation in Africa. Sea-level rise, changes in precipitation and temperature and the high frequency of extreme weather events will be the main features of climate change. They will probably have adverse impacts on the natural resources, the coastal systems, the economic activities and on human health in Sub-Saharan Africa.

5.2.1 Climate Change and Natural Resources Management

5.2.1.1 Water Resources

Independently of considerations related to climate change, a growing number of African countries are in a situation of water stress or scarcity resulting from a combination of factors including decrease in precipitation, destruction of catchment basins and population growth. Within the limits of the arid and semi-arid areas, the Sahelian region for instance, the image of the African woman covering distances of several kilometres in a daily ritual for water supply has become an ordinary fact. Yet that problem, which is at one and the same time harmful for her health and time consuming, could be exacerbated by climate change and expanded over wider scales. This would mainly result from a possible decrease in rainfall, which in addition will reduce the runoff in different river basins and globally affect the hydrological balance of the continent (IPCC 1998). As a consequence of recurrent drought, Lake Chad has shrunk from 23,500 km^2 to 2000 km^2 within less than three decades (Suliman 1990). This is an illustrative case that should be meditated because a similar phenomenon is expected to affect some African rivers; to bring about the spread of diseases associated with water scarcity and to disturb the functioning of dams.

5.2.1.2 Ecosystems

A few African terrestrial ecosystems are spared by human pressure or climate instability. Even, certain protected areas such as forests do not escape the threats. A lot of care is needed to defend their physical integrity against various types of poaching and agricultural expansion which is generally the most pressing threat. The prevalence of unsustainable management practices and the lack of proper judgement or long-term vision lead to an immoderate consumption of resources. This is naturally prejudicial to the ecosystems.

In a context where the threats correspond to changes in the climate parameters, the flora and fauna's response to the changing conditions take place over long periods of time. If they have the capability to survive, they have either to adjust themselves to the new conditions or to shift to more favourable latitudes or altitudes. However, that response is not always feasible for all species or organisms since it may be impeded by the rapidity of the changes. According to the IPCC (1998), such a phenomenon is likely to happen in Africa. It will disrupt the balances within most of the ecosystems, especially forests and rangelands/grasslands. Almost all the regions will be concerned in particular in the arid zones. An increase of temperature within a few decades associated with a decrease of rainfall will certainly annihilate the response capabilities of some vegetal and animal species and compromise their survival.

But in certain areas in Eastern and Central Africa where climate change is expected to have positive impacts by altering favourably the rainfall patterns and subsequently enabling the revival of marginal lands, more and more people are likely to be attracted. Through their usual unsustainable practices, these people could rapidly exceed the carrying capacity of the lands and thus misuse or reduce to nothing the mercy of climate change (IPCC 1998).

5.2.2 Climate Change and Economic Activities

All the economic activities that depend on natural resources will be affected by the consequences of climate change. In the context of Sub-Saharan Africa, these are known to be mainly agriculture and fisheries. They both have a critical importance for the survival of thousands or even millions of people. Besides, they are in most countries the principal sectors that create wealth because of their incomparable contribution to Gross Domestic Products.

The consequences of climate change on the Sub-Saharan African agriculture and food security will be devastating. The current situation is already pathetic. Intraseasonal and interannual variability of rainfall, demographic growth and land degradation combine their effects to engender severe disruptions to regional food supplies. Irrigation is not a well-developed practice although it can be used as a buffer against climate variability. It concerns an insignificant proportion of the cultivated lands. Consequently, rain-fed agriculture is the dominant activity and its productivity is generally affected by the "caprices" of the climate. This is likely to worsen because warmer temperatures may induce a rise in potential evapotranspiration that a hypothetical increase in rainfall will not necessarily cover. In addition, the alteration of variables such as temperature, wind and humidity by climate change will certainly contribute to the spread of pests and diseases detrimental to a normal growth of crops (Rosenzweig and Hillel 1998). So, there is a considerable risk of exacerbation of the current African food crisis.
Without doubt, the disaster will be greater if the impacts of climate change on fisheries lead to a reduction of the fish landings. This is regarded as a possible result of changes in water temperature and in the pattern of stream flows. If migrations are feasible for marine

47

fisheries, this possibility is naturally limited for freshwater fisheries because of the small size of the biota (IPCC 1998). In some African countries where the principle of biological rest is not properly applied, the artisanal fisheries are becoming less and less productive because of overfishing. The perspectives of climate change will maybe further deprive certain African populations of an important source of proteins and income. In the same way, countries may suffer because fisheries are an important source of employment and a major sector for exports.

A third sector similarly dependent on natural resources and emerging progressively as one of the most promising industries in Africa is tourism. Considered as the last bastion for many great felines and other endangered mammals, the Sub-Saharan region abounds in sites and reserves that provide a growing number of visitors direct contact with or a rediscovering of wildlife. Even though Africa is currently playing a minor role in the world tourism, the expansion of the tourism industry is constant and the potential is still considerable. In 1998, it received 4% of the 626 million international tourist arrivals and 2.1% of the $445 billion generated by the activity (The Courier n° 175 1999). But how can the future of this industry in general and that of ecotourism in particular be guaranteed if the Sahel as well as Eastern and Southern Africa are expected to experience repetitive droughts? In a nature reserve like "Serengeti" in Tanzania, it is touching to see the heavy tribute paid by the herds of gnus during each dry period and the direct consequences on their own predators. Maybe more violent emotions are waiting for us in the coming decades.

Globally, the economic activities as well as the ecosystems are placed under the threat of another possible outcome of climate change, which is a high frequency of extreme weather events. In the last decade, different parts of the world have been victims of a recrudescence of extreme weather events such as torrential rains followed up by flooding (Myers and Kent 1995). The most recent cases concern Venezuela, Mozambique and Brazil. Probably, the increase of evaporation made possible by warmer temperatures entails a high frequency of these tragic events. In the case of Mozambique considered as one of the poorest countries of the world and which is always suffering from the effects of a long civil war, the damages and the losses have been considerable. An assessment carried out by the government's disaster management agency reveals that the floods have cost US $45 million in lost exports, US $245 million in lost production and US $270 million of damage to property in both public and private sectors (Archives Reliefweb August 2000). Naturally, these estimates do not include the dead persons and the outbreak of diseases.

5.2.3 Climate Change and Public Health

As indicated in subsection 5.1.2 "Poverty, Malnutrition, Diseases and Public Health", the geographical distribution of infectious diseases world-wide is clearly at the disadvantage of Sub-Saharan Africa. The cost paid by the region is enormously high because every year, the morbidity and mortality rates associated with these diseases concern millions of people. Manifestly, the subcontinent disposes of climate conditions favourable to the prevalence and the spread of water-related and vector-borne diseases such as diarrhoea, cholera and malaria. Temperature, precipitation, humidity and wind are the main climate variables whose fluctuations contribute to their spread (WHO 1996). Whereas the current situation presents the features of a disaster that affects nearly the whole region, the incidence of vector-borne diseases is likely to increase again under the influence of climate change. That will be the case of potential malaria risk expected to increase world-wide. A warmer environment could extend abundance of mosquitoes and open up new areas of malaria. In that perspective, the prevalence of malaria will probably increase in the coming decades to cover the whole continent and it might gain ground in some areas of Southern and Eastern Africa where its endemicity is currently low.

This spread of malaria has maybe already started because in the highland regions of Ethiopia and Kenya, which were previously safe from the disease, new cases are being increasingly reported. Yellow fever seems to follow the same trend in the same areas. The outbreak of infectious diseases, especially water related diseases will also result from the high frequency of extreme weather events such as floods and droughts. A simple direct exposure to flood for instance is likely to cause injuries and deaths. In case of drought, the scarcity of water increases the concentration of pathogenic organisms in raw water supply and consequently favours the emergence of cholera, dysentery and other diseases. Furthermore, crop failures will deteriorate the nutritional status of communities and will increase their vulnerability to epidemics (Otieno 1999).

So the potential health effects of climate change on Sub-Saharan Africa will certainly lead to tragic situations for the populations. Most of the governments that are easily overwhelmed by the magnitude of the current problems will have to deal with a higher incidence of diseases and extreme situations.

5.2.4 Impacts of Sea Level Rise on Human settlements

The most important African cities are generally located on the coastal areas and benefit from an outstanding concentration of populations, infrastructural facilities and economic activities. This seems to be part of the colonial legacy as the logic of resources exploitation during colonial rule favoured to a large extent the construction of ports and the emergence of human settlements along the African coasts. The perspectives of thermal expansion of the oceans are raising some concerns about the future of these areas. Very likely, in the coming decades, they will have to face the rising challenge of the sea. The most vulnerable countries have been identified as those with low-lying coasts, which are currently suffering from storm surge and coastal erosion. Such countries can be found in the different regions of the continent. But they seem to be more numerous in West and Central Africa (IPCC 1998).

In a context of low-lying coasts, the direct consequences of a local increase in the sea level will be the erosion of the shorelines and the flooding of the surrounding areas, which can consist of agricultural lands, industrial infrastructures or human settlements. Besides, such an event is likely to favour the intrusion of salt-water into rivers on distances going beyond the estuaries and the deterioration of the quality of the aquifers (Hulme et al 1995).

Naturally, the extent of the disaster will be a function of the magnitude of the sea level increase. One-meter rise in sea level is a hypothesis that is often used in the studies designed to assess the potential effects of climate change on the African coastal systems.

Generally, these studies lead to conclusions that thousands of km^2 of land will be eroded and/or flooded entailing considerable economic losses and thousands or millions of people will necessarily have to be relocated. In this respect, the content of the Box 5.2 presents the risks incurred by Senegal and Nigeria in case of one-meter rise in sea level. These are illustrative and representative of what is likely to affect numerous African countries.

Box 5.2: Impacts of Sea Level Rise in Senegal and in Nigeria

- *The Senegalese coastal zone extends over 12,150 km^2 and represents about 3% of the national territory. It accommodates about 5 million people (2/3 of the country's population) and concentrates most of the companies (90%). Mainly constituted of sandy lands surrounded by some dunes, it is currently subjected to erosion that sets the shoreline back in the range of 1.25 – 1.30m per year. At a regular rhythm, some edifices are destroyed and the rivers are gradually invaded by seawater leading to an increase in salinity from the mouth to upper waters. This fragile situation will be exacerbated by a one-meter rise in sea level. With such an event, 6042-6073 km^2 corresponding to about one half of the coastal zones will be lost through flooding and erosion; 109,000-178,000 people will be obliged to leave their homes for a relocation in other areas and the assets at risk represents a value of US$499-707millions.*

- *In Nigeria, a similar increase in sea level will lead to more destructive impacts. Much of the population and economic activity, particularly the oil industry, is located along the coastline. The estimates have revealed that one-meter rise will probably cause the flooding of 18,000 km^2 of land, the displacement of 3,680,000 people and damages on assets valued at US$9 billion.*

Source: IPCC 1992; MEPN 1997

Among the impacts of climate change projected for Sub-Saharan Africa, the consequences of sea level rise are likely to be the only unusual problems in the context of the subcontinent. Most of the consequences correspond to problems that are more or less well known and established. However, the general opinion seems to indicate that climate change will add a supplementary dimension to the current difficulties. The extent and the frequency of the expected problems do not have any precedent in the history of the region. All aspects of people's lives will be concerned by the spectrum of impacts described above. The future of the region is really at stake.

While most countries of Sub-Saharan Africa are already stretched to their survival limits, their fragile natural resource base and the physical integrity of their populations are placed under additional threats. Consequently, climate change is a phenomenon that clearly unites the problems of the environment and those of development. The vulnerability of Sub-Saharan Africa to the potential effects of a changing climate is a fact unanimously recognised by scientists and the observers. However, it would be naïve to believe that such vulnerability results only from the recurrent climate variability. The deep causes are rather the economic destitution and the social distress of most African countries. They are desperately trying to come to grips with their financial, technological and institutional deficiencies. In the absence of positive economic and social development, it is clear that Sub-Saharan Africa is not likely to successfully manage the effects of climate change. Taking up the challenge of climate change in Africa cannot and should not be envisaged without a reconciliation between environment and development. As long as development is promoted at the detriment of the environment, the vulnerability to climate change will increase. Hence, it is essential to establish the notion of sustainability as the common denominator of all the development policies in Africa and to prepare the subcontinent to effectively manage the consequences of climate change.

Chapter Six
Meeting the Challenge of Climate Change in Sub-Saharan Africa

Obviously, taking up the challenge of Climate Change in Sub-Saharan Africa is an ambition that cannot be achieved in isolation or in rupture with the efforts made at a wider scale by the international community. Through the United Nations Framework Convention on Climate Change (UNFCCC) and the Kyoto Protocol, an international regime is being progressively set up and consolidated in order to tackle an issue that concerns the whole planet. As most other international regimes, this one should provide the opportunity to elaborate an adequate response to a pressing threat. To deal with the climate change problems implies averting the risks, and if not possible, mitigating the potential effects through the adaptation of appropriate measures. Naturally, the success of this collective enterprise depends on the contribution of each country. But, such contribution cannot be the same for all since global warming is largely caused by industrial activities in the developed countries. Hence, they have a historical responsibility ensuring the success of the global response devised by the international community. In other words, there should be less tergiversation from the developed world and more commitment for the reduction of the greenhouse gas emissions. This necessarily implies changes in people's lifestyles and in the current model of development. Without a resolute commitment of the industrialised countries, the efforts undertaken in other regions such as Sub-Saharan Africa in order to cope with climate change would be in vain and meaningless.

For a successful combat against the challenge of climate change in Sub-Saharan Africa, priority has to be given to two areas of actions. First, owing to the global character of the issue, it is essential that the international community strives to ensure the achievement of the ultimate objective of the Framework Convention of Climate Change. There should be a perfect adequacy between the will to prevent dangerous human interference with the climate system and the commitment to limit and reduce greenhouse gas emissions.

Second, it is urgent that the countries of Sub-Saharan Africa combat vulnerability through the adoption of development policies consistent with the restoration and the protection of their natural resource base.

6.1 Improving the Consistence of the Commitments under the Kyoto Protocol

The Kyoto Protocol has been added to the international regime on climate change because the commitments existing under the UNFCCC were considered insufficient to bring significant changes in the climate policies of most industrialised countries. However it is still to be known whether the provisions included in the Protocol, which mainly consist of legally binding obligations to limit or reduce the major greenhouse gases responsible for global warming are adequate.

6.1.1 Stabilisation of Atmospheric Concentrations of Greenhouse Gases

As presented in Box 3.2, the convention seeks to stabilise the atmospheric concentrations of greenhouse gases at a level that would prevent dangerous human interference with the climate system. The time allowed for reaching that level is the required duration to enable the natural adjustment of the ecosystems, the elimination of threats over food production and the strengthening of the prospects of sustainable economic development.

An overall reduction by at least 5% in the emissions of six greenhouse gases including carbon dioxide, methane, nitrous oxide, sulphur hexafluoride, hydrofluorocarbons and perfluorocarbons has been consented to by the industrialised countries in the commitment period from 2008 to 2012. It is assumed that this "six-gases approach" also called "basket approach" will provide each Annex B Party with more flexibility by enabling it to determine its reduction efforts on the basis of its specific emission profile. But what is really the level of significance of this commitment?

The first certainty is that it is really far from having the same consistence as some reduction proposals made during the negotiations by certain groups of countries. Most of these were based on a different approach with a gas-by-gas reduction target and a gradual evolution over time. For instance, early in the negotiations, the Alliance Of Small Island States (AOSIS) suggested the adoption of target for each major GHG and in particular a 20% reduction in the emissions of carbon dioxide by the year 2005 (Oberthur and Ott 1999).

Moreover, if we only refer to the 6 billion tonnes of carbon added each year to the atmosphere from fossil fuels and other industrial sources (Nilsson 1992), the 5% reduction of six GHGs mentioned above will appear insignificant and even ridiculous. For sure, it will not enable a rapid stabilisation of the atmospheric concentration of the main GHGs. A long time will be needed to achieve the ultimate objective of the UNFCCC unless the industrialised countries accept to reduce more significantly their GHGs emissions.

In its second assessment report, the IPCC considers that the immediate stabilisation of atmospheric carbon dioxide concentration at its present level requires an immediate reduction in its emissions of 50 to 70% and further reductions thereafter. Some think that the objective of the convention does not stress or underline enough the urgent or alarming aspect of the present situation. The idea of stabilisation of the atmospheric concentrations of greenhouse gases is discussed in a context where it may be more relevant or more adequate to go beyond a simple stabilisation. In fact, the emissions of GHGs should be drastically cut. The basket approach with its proposed minor reductions of the GHGs emissions may be encouraging for the future, but cannot lead to a rapid and substantial decrease in the accumulation rate of the main gases responsible for global warming. The decision-makers are probably encountering difficulties to determine the level of atmospheric concentrations of GHGs that might interfere dangerously with the climate system. That is not an easy task because a lot of ecosystems with different sensitivity[8] to climate change are subjected to threats. Thus, under the influence of some uncertainties and the pressure of certain economic circles, the international community seems to proceed by trial and error. That attitude is likely to last because of the difficulties to

---

[8] Sensitivity is the degree to which a system will respond to a change in climatic conditions (e.g. the extent of change in ecosystem composition, structure and functioning resulting from a given change in temperature or precipitation) (IPCC 1995).

rapidly find alternatives to fossil fuels-based technologies and the reluctance of certain countries to change the direction of development. Yet, the climate risk should not be regarded as a minor problem since the future of many countries is considered to be at stake.

6.1.2 Role of Sub-Saharan Africa

Sub-Saharan Africa is a vulnerable region with regard to climate change. If the international community fails to modify the climate policies applied in the industrialised countries, the potential for climate disturbances will certainly increase and so will the risks of detrimental effects on Sub-Saharan Africa. All initiatives that the international community can take to prevent such a situation would be salutary. Also lethargy and inaction within the region would be equally suicidal. In both debate and actions concerning climate change, Sub-Saharan has a vital role to play. The climate change issue provides the region with an excellent opportunity to reconsider its social and economic development policies and practices.

6.2 Combating Vulnerability to Climate Change in Sub-Saharan Africa

Because of the low levels of industrial development and economic growth in Sub-Saharan Africa, the GHG emissions in the region are comparatively insignificant. They amount to less than 7% of global emissions and to about 4% of carbon dioxide emissions (Sokona, Humphrey and Thomas 1998). Clearly, the subcontinent of Sub-Saharan Africa has only a minor contribution to the emergence of the global warming problem. For this reason, it is not for the moment compelled to reduce its greenhouse gas emissions. Consequently, the challenge for the policy-makers in each country of the region is not to focus on GHG mitigation options, which are very limited but to find the appropriate policies for combating the vulnerability to climate change. In this perspective, it would not be appropriate to hesitate in the choice between the two strategies of "No Regrets" and "Wait and See", which are at the heart of the debate over policies response to climate change. It is obvious that Sub-Saharan Africa needs to adopt the "No Regrets" strategy since climate change is not expected to bring about completely new problems. It will simply exacerbate existing problems.

Furthermore, it can be assumed that the veritable challenge for Sub-Saharan Africa is the search for development take off. Its extreme vulnerability to climate change, which is unanimously recognised by the specialists, results more from the lack of institutional, economic and social capacities than from an incurable predisposition to be hit by climate instability. Without good development perspectives in the region, it would be illusive to expect some achievements in the attempts to cope with climate change and to deal with its associated problems.

6.2.1 Realigning Development Policies.

There should not be dissociation between the development objectives and the specific actions taken in order to anticipate on the expected impacts of climate change. On the contrary, as reasonably suggested by Sokona, Humphreys and Thomas (1998), the climate change issue in Africa has to be resolutely placed in the context of the global development policies. However, it would not be realistic and easy to convince the African leaders to consider climate change as their first priority. The daily lives of the population are so hard that the hypothetical disasters that threaten the future might not be taken into consideration. Yet, it is critical to find a compromise between the current social, economic and political urgencies and the challenges of the future. A similar balance is also to be established in the relationship between human needs and resource availability of the Earth in order to pursue people's welfare within the natural environment capacity (WWF 1993).

Which model of development is able to achieve such a compromise? The current models of development, which are mainly designed to produce economic growth, certainly do not facilitate such a compromise. They have lamentably failed to integrate environment and development. The hopes are now resolutely oriented towards the sustainable development approach. Even though it is likely to remain an eternal challenge, the efforts made in favour of its promotion might bring the salute that African countries crucially need. It possesses the rare virtue of being concerned about the conservation of the natural resource base and to have a similar consideration for both present and future generations.

Considered under the perspective of climate change, it could help degraded areas to restore their natural resource base through environmental rehabilitation. This is a crucial step in the context of Sub-Saharan Africa where the natural resources are subjected to a drastic depletion. According to Suliman (1990), it is a fundamental prerequisite for averting the negative implications of climate change and ensuring economic recovery and transformation.

Naturally, if the development objectives are to be associated with the imperative of an effective management of climate change, an integrated approach grounded on the principles of sustainable development is needed. In most of the key areas identified in chapter 5 as being potentially vulnerable to climate change, the management practices have to be reconsidered so as to reconstitute the natural capital and to reverse the unsustainable trends.

However, the most pressing urgencies certainly concern the two sectors of agriculture and energy. Agriculture is the main economic sector in Sub-Saharan Africa since it generates 20-30% of the Gross Domestic Product (GDP) in most countries (IPCC 1998). Energy consumption is considered to be incredibly low in Africa, in comparison with the situation in the developed world; on average, African population consumes 30 times less energy than North American people (Cissé, Sokona and Thomas 1997). There is a strong need to increase energy consumption in order to promote economic development. In Sub-Saharan Africa, it is known that both sectors contribute the most to making the region vulnerable to climate change. People's propensity to expand the agricultural fields at the detriment of the forests and to meet their energy needs mainly with biomass leads to a high rate of deforestation, which annually strips a surface more than three times larger than the republic of Gambia. These practices put more and more land to degradation, cause a continuous decrease in agricultural yields and increase the destitution of the rural populations. Thus, a dramatic vicious circle is created in which poverty leads to environmental degradation and in turn, environmental degradation spreads poverty. Breaking this cycle would be a decisive step; it is a critical requirement of any policy designed to promote environmental rehabilitation and socio-economic development in Sub-Saharan Africa.

6.2.2 Rehabilitating the terrestrial ecosystems.

Degraded resources are unlikely to successfully adjust to changing conditions. The conservation or the protection of the soils, forests and wildlife against recurrent drought and mismanagement practices is to be considered as a preparedness element in anticipation of climate change. These practices consist mainly of unconsidered agricultural expansion, suppression of fallow periods, anarchic deforestation, etc. They have to be reconsidered in order to reduce land degradation and promote a sustainable use of the resources.

The implementation of a combination of options would be necessary to get to that end. These include a decrease of the rate of tree cutting through deforestation control, maintenance and improvement of the genetic diversity of the forests through the support to silvicultural research and constitution of seed banks, combating erosion, maintenance and expansion of the nature reserves. The adoption of participatory approach to implement these options may significantly contribute to success.

6.2.3 Developing Alternative Sources of Energy.

With regard to the energy sector, it is perfectly legitimate to claim an increased consumption in Sub-Saharan in response to the requirements of socio-economic development. Also, the current patterns of energy use have to be radically modified. It is no more acceptable that biomass remains the main source of energy in most African countries. In 1990, biomass represented more than 80% of the total energy used in Sub-Saharan Africa. For the current year, the figure is likely to be the same (WRI et al 1998). Naturally, biomass is mainly provided by the forest ecosystems, which are rapidly disappearing and causing the loss of biodiversity and the degradation of soils. It is very urgent to slow down the deforestation rate, to initiate more reforestation programmes and to find new substitutes to replace the systematic use of wood and charcoal. A first step in this direction would be the improvement of the efficiency of energy use in industries as

59

well as in households. For the long term, it is crucial to favour the development of renewable sources of energy, especially biogas and wind and solar energy. The potential exists and this will progressively decrease the pressure on the forests, help to maintain barriers against soil erosion and biodiversity losses and prevent the destruction of carbon sinks.

6.2.4 Increasing Agricultural Production.

The challenge is to devise a model of agricultural development that would be productive enough to keep pace with demographic growth without entailing any environmental damage at the detriment of the forested capital. In this respect, the association between agriculture and forestry and the use of safer and cheaper fertilisers (for instance the valorisation of organic waste as fertilisers) provide some interesting perspectives for the protection of the soils, the increase in the yields and the creation and/or maintenance of carbon sinks. In addition, for crops as well as for forests, it is crucial to develop more drought tolerant varieties and to improve the efficiency of irrigation whose current potential according to World Bank (1996) is substantially under-utilised.

Besides, a sustainable practice of animal husbandry requires the maintenance of an adequate livestock population in order to avoid the phenomenon of overgrazing detrimental to the rangelands.

6.2.5 Improving the Management of Water Resources.

It is quite easy to agree on the necessity to find as quickly as possible the ways to improve the management of the water resources in Sub-Saharan Africa. A growing number of countries in the subcontinent are either already suffering from water scarcity or are under its pressing threat. Generally, recurrent droughts, population growth, increase of livestock and unsustainable uses are the main contributory factors. As mentioned in subsection 5.2.1.1, this situation is likely to be exacerbated by climate change. Some of its possible outcomes, for instance decrease in rainfall or sea level rise, may have detrimental effects on water supplies through the alteration of both quantity and quality of freshwater. This will accentuate or bring about water shortages, which in turn will affect the environment, food production, human health and economic growth.

In response to these worrying perspectives, it is imperative to improve the water management policies and practices by making them more sustainable. This implies a more efficient use of water at all levels of the society, the protection of the watersheds and the generalisation of the practices of catchment and water harvesting during the rainy seasons.

In a city like Dakar where the groundwater reservoirs are overexploited and threatened by saltwater intrusion (WHO 1996), an important part of the rainwater is collected through a network of channels and discharged into the surrounding seas. This system is naturally inefficient and absurd because it entails the loss of a scarce resource that could be used to recharge the aquifers. In water-stressed countries, especially in the cities, it is urgent to design the rainwater management system so as to harvest as much water as possible during the rainy seasons.

6.2.6 Promoting a Sustainable Management of the Fisheries Resources.

In addition to be an important source of proteins for the African populations, the fisheries are also an important sector for employment, exports and creation of wealth. But they incur some risks of disturbances, which might result from changes in water temperature, modification of pattern of stream flows, deterioration of nursery areas, etc. The African countries can respond to these risks, which have the high potential to jeopardise the fisheries resources, by:

- The restoration and the implementation of protective measures on the nursery areas,
- The promotion of aquaculture,
- The prohibition of overfishing and reduction of the activities during the periods of reproduction,

- The refinement of the fishing techniques to make them consistent with a sustainable use of the resources (control of the fishing methods and prohibition of the use of dynamite).

6.2.7 Protecting the Coastal Regions.

The presence of particular ecosystems and the importance of the socio-economic systems in the African coastal areas justify the urgency of implementing strategies for adapting to possible sea level rise. To adequately define the type and the timing of the responses, each country will have to assess the vulnerability of existing activities and to determine the magnitude of the risks. Since various sectors and interests are represented in the coastal zones, a large framework is needed to achieve such assessment and a broad planning and management context to implement the response strategies. Among the adaptation measures, constructions of small or large dikes, relocation of socio-economic facilities, displacement of populations and changes in land use are often evoked. Most of them may be very costly because of the size of the African economies.

However, beyond the protection of the coastal areas by dikes or other seawalls, it would be also appropriate to maintain a strict control over the high-risk zones and to discourage the establishment of new investments in these vulnerable sites. Furthermore, there is strong need to promote regional development in most countries of Sub-Saharan Africa. The concentration of the most important economic infrastructures in the coastal regions contributes to the depopulation of the hinterland and brings about a global imbalance in the distribution of the population. This increases the vulnerability of many national economies to sea level rise. Valorising the potential existing in the hinterland would open new opportunities for public and private investments and would certainly help to slow down the rural migrations.

6.2.8 Responding to the Threats on Public Health.

The present state of public health in most countries of Sub-Saharan Africa is not satisfactory. It needs to be considerably improved because the tribute paid to diseases, particularly to infectious diseases, amounts to millions of victims per year. Many social and environmental factors such as undernutrition, poor sanitation practices and deterioration of the quality of the environment are involved or associated with the bad health status of people. It is believed that climate change will make the burden heavy by extending the areas of risks and by increasing the number of infections. Besides, injuries and outbreaks of diseases are likely to occur in areas hit by flood and other types of extreme weather events.

Beyond the need to strengthen the health care services, the adoption of a preventive approach would much help to improve the current situation and to deal with the risks associated with climate change. In this regard, the following measures are of crucial importance:

- Providing clean and potable water to rural and urban populations
- Improving the sanitation[9] conditions
- Intensifying vaccination programmes
- Initiating awareness raising programmes on environment and health issues

- Setting warning systems up to minimise loss of life, damage to property and disruption of food supplies in case of natural disasters.

The sectoral adaptive measures to climate change presented in these subsections do not represent standard solutions, which have to be blindly applied on all local or national situations in Africa. They rather constitute some broad options that should be more investigated and refined by the countries of Sub-Saharan Africa when trying to define their climate policies. The adoption of one adaptive measure rather than another should be decided on the basis of the specificity of each local or national context.

The sectoral approach only provides partial ways of dealing with climate change. It is not designed to identify the interactions between the different dimensions of the issue. This is not satisfactory because the response to climate change is to be global, coherent and

---

[9] Sanitation refers to all conditions that affect health, especially with regard to dirt and infection and specifically to the drainage and disposal of sewage and refuse from houses (Franceys, R., J. Pickford and R. Reed 1992).

visible in the whole development policy. Hence, a step further has to be taken in order to get a global picture of the possible outcomes of climate change. Such a picture can be provided by an integrated vulnerability assessment to climate change. This will enable each country to catch all the cause-effect relationships from human-induced climate change at global level to the repercussions on the daily lives of people at local scales and to identify the interactions between natural, social and economic systems.

Climate Change: possible impacts and response at regional, national and local scales

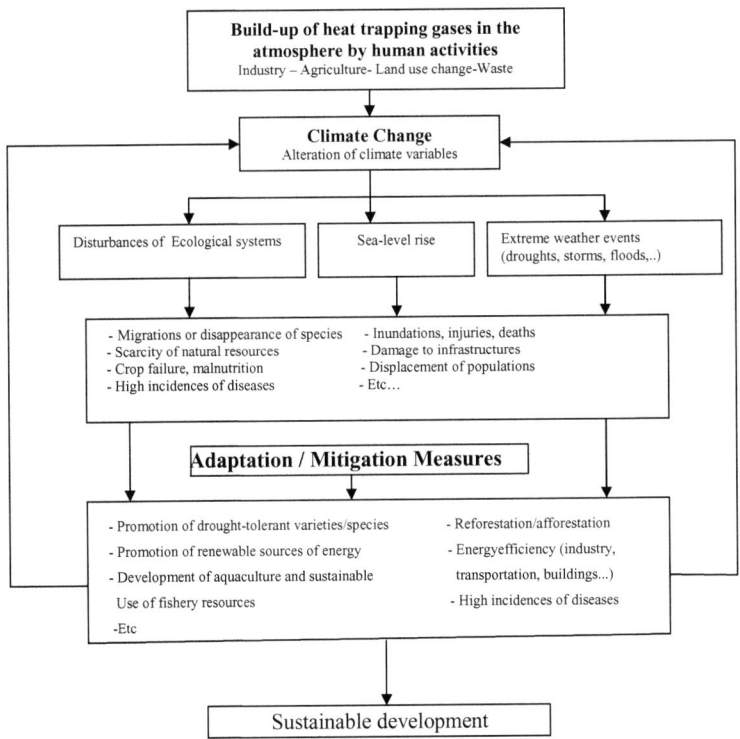

Chapter Seven
Conclusion and Recommendations

Climate change is assuredly a perspective that should not be considered with nonchalance and unselfconsciousness. Some scientific uncertainties still exist, but these do not suppress the reality of the threat. This has been understood by the international community, which seems to have now a common global view on the determinants of the problem. Human activities, especially industrial activities, have been releasing for two centuries more and more heat-trapping gases, thereby modifying the chemical composition of the atmosphere. This is leading to disturbances in the energy balance of the Earth-atmosphere system causing a gradual warming of the terrestrial atmosphere. From this relationship has emerged the fear of a climatic change.

Naturally, this view is much debated. But the contesting seems to be at present less virulent. It is particularly persistent among the circles that defend specific economic interests connected with the coal and oil industries. The work carried out within the Intergovernmental Panel on Climate Change (IPCC) has certainly contributed to draw together the scientists' views and favoured the achievement of a wider consensus. Besides, the multiplication of extreme weather events throughout the world during these last decades, under the form of floods, drought and others, has maybe produced some psychological effects and persuaded a majority of people that climate change is on the way.

The international community seems to be no more setting out in search of scientific motivations for a global response to climate change. Defining the modalities of an international action is presently the main concern. It is at the heart of the debate within the international regime on climate change. Three flexibility mechanisms known as Emission Trading (ET), Joint Implementation (JI) and Clean Development Mechanism (CDM) have been outlined and included in the Kyoto Protocol. They are mainly designed to enable countries subjected to "quantified emission limitation or reduction objectives" (QUELROS) to achieve their obligations at lower costs.

Globally, the industrialised countries are committed to reduce by a least 5% their emissions of six major greenhouse gases in the period 2008-2012. It is difficult to believe that this agreement is satisfactory, especially if one considers that it will not substantially slow down the build-up of greenhouse gases in the atmosphere. Among these, carbon dioxide has the largest contribution to the enhanced greenhouse effect. Annually, 26 billion tonnes of this gas are released by human activity. About half of this amount remains in the atmosphere while the second half is believed to be absorbed into the oceans and by vegetation on land.

How can the build-up of greenhouse gases in the atmosphere be stabilised or even reversed? The objective of the United Nations Framework Convention on Climate Change (UNFCCC) is clearly to stabilise atmospheric concentrations of GHGs at a level that would prevent dangerous human interference with the climatic system. The time required for such a stabilisation is not determined with precision. It should correspond to the time necessary for the achievement of three crucial things: the natural adjustment of the ecosystems, the preservation of world food security and the promotion of sustainable economic development. This situation provides the Parties to the UNFCCC with some kind of latitude or flexibility. The first commitments accepted by the industrialised countries under the Kyoto Protocol, a 5% reduction in the global emissions of carbon dioxide, methane, nitrous oxide, sulphur hexafluoride, hydrofluocarbons and perfluorocarbons, sound like a symbolic decision. They do not reveal a decisive will to rapidly evolve towards the achievement of the ultimate objective of the UNFCCC. Hence, the risk of climate disturbance is likely to keep on growing and the threats over the vulnerable countries being more worrying.

Pointed out as one of the most vulnerable regions, Sub-Saharan Africa is likely to be subjected in the coming decades to serious afflictions because of climate change. Distressing events such as repetitive droughts, famine, epidemics and floods have featured its past and recent history. It is feared that the impacts of climate change will worsen the current situation, mainly through the alteration of the precipitation patterns, the multiplication of extreme weather events and the exacerbation of the stress over the natural resource base. Even though it is impossible to indicate with precision the extent of the coming disasters, the future of the subcontinent is believed to be at stake. Each aspect of people's lives may be affected and agriculture, the backbone of the African economies, is probably the sector where the risk is higher.

For the subcontinent, it would certainly be suicidal to ignore the threats and to adopt a "wait-and-see" strategy. The current situation in many respects is close to a disaster. Recurrent droughts and unsustainable resource use are leading to a continuous shrinking of the natural resource base. Poverty, undernutrition and poor health status are spreading and becoming the features of the living conditions of a growing number of people. Such a situation is more than a burden. It sounds like a nightmare and explains why Sub-Saharan Africa is believed to be extremely vulnerable to climate change. Inaction would be unreasonable. It is rather vital for this region to strive to take up the challenges of climate change, which is very likely to exacerbate the existing problems.

In this perspective, combating or reducing vulnerability to climate change is of paramount importance. This requires profound and radical changes in the current development policies and practices, which are manifestly failing to pursue human welfare within the carrying capacity of supporting ecosystems. The countries of Sub-Saharan Africa have to adopt a new development approach that can confront the current problems and provide the future generations with the means to take up the challenges of their times. In other words, these countries would do well to adopt the sustainable development model.

Of course, promoting a "process of change in which the exploitation of resources, the direction of investments, the orientation of technological development, and institutional change are made consistent with future as well as present needs" (Boon and Hens 1994) is an immense challenge. It requires the acquisition of appropriate capacities at all the levels of the African societies, from States to organisations, communities and individuals. The States in particular are expected to have the capabilities to design and implement plans and policies aimed at achieving sustainable development. They have to set up an adequate legal and institutional framework that enables a correct handling of the development problems and provides individuals, communities and organisations with the opportunities to acquire skills and to get involved into the development process in a positive way.

The African countries and the whole developing world are not subjected to legally binding obligations to reduce their emissions of greenhouse gases. This is because they have not significantly contributed to the emergence of the global warming phenomenon. Also, they are not required to participate in the global effort of reduction of greenhouse gases emissions in order to avoid the addition of new constraints to their development process. However, this position does not exempt them from initiating voluntary actions. Within the international regime on climate change, an agitation is insidiously developing under the influence of the United States of America calling for more involvement from the Southern countries. This agitation seems to target in particular countries like China. But the countries of Sub-Saharan Africa may be more and more heckled in the polemic debate about the alarming destruction of the tropical forests. The protection of these forests has a crucial importance. It is an important aspect of the world effort designed to prevent or reduce the loss of biodiversity and an essential dimension of the maintenance of terrestrial sinks of greenhouse gases. That is why the preservation of the tropical forests can be legitimately regarded as part of the response to the climate change issue. Trying to have an international legislation on the tropical forests invariably leads to a determined opposition of leader countries in Africa and Asia. Yet, the temptation is growing within certain international organisations or in particular industrialised countries to associate their support to the developing countries with some obligations relative to the protection of the environment. The African countries would do well to spontaneously protect their natural capital and to promote the use of environmentally-sound technologies. This is critical for their prospects of development. In the coming times, they may be required to consent to some commitments under the Kyoto Protocol in exchange for the assistance they crucially need for strengthening their capacities, dealing with poverty and improving control over their destiny.

In consequence, the countries of Sub-Saharan Africa will be advantageously prepared in dealing with the expected impacts of climate change and in pursuing sustainable development paths if the recommendations provided below are included in the development policies and in the sectoral strategies, with a status of priority objectives.

General Recommendations

- Protection of the natural resource base, specifically soil, water, vegetal cover and wildlife against further destruction; rehabilitation of the degraded ecosystems.

- Encourage regional development in order to contribute to a larger distribution of the economic facilities throughout each national territory and to reduce the vulnerability of each country as a whole.

- Promotion of economic diversification to increase the potential and enlarge the supports for the perspectives of development.

- Encourage and support more research on development issues, particularly on the social and economic sectors such as food security and water supply, which are considered to be sensitive to climate change, with a view to conduct national and local vulnerability assessment studies and to verify the adaptability[10] of the response strategies to specific local and national conditions.

- Exchange of information and experiences between countries of Sub-Saharan Africa and promotion of regional integration.

---

[10] Adaptability refers to the degree to which adjustments are possible in practices, processes or structures of systems to projected or actual changes of climate (IPCC 1995).

Specific Recommendations

- Initiation of afforestation programmes to extend the forested areas with the use of productive and drought-tolerant species.
- Development of means for the prevention of bushfires.
- Adjustment of the demand of woodfuel to the real potential of the forests.
- Prohibition or limitation of the imports of second-hand vehicles and promotion of public transportation.
- Control of air pollution and creation of warning systems.
- Improvement of energy efficiency in the industrial processes and adoption of environmentally-sound technologies.
- Improvement of energy efficiency in the buildings designed for housing, administration, commerce, etc.
- Development and promotion of renewable energies, solar and wind energy in particular.
- Recovery and valorisation of the methane naturally produced in the landfills.

The adoption of these recommendations, according to the specific conditions of each country, may help Sub-Saharan Africa to associate the pursuit of sustainable development objectives with a responsible participation in the global effort designed to deal with the climate change issue.

Bibliography

Abbott, P.L. 1996. *Natural Disasters*. Dubuque (Iowa): Wm. C. Brown Publishers.

Adams W. M. 1990. *Green Development: Environment and Sustainability in the Third World*; London and New York: Routledge.

Agnew, C. T. 1995. Desertification, drought and development in the Sahel. In *People and Environment in Africa*, edited by Tony Binns. Chichester: John Wiley & Sons Ltd.

Archives Reliefweb. 2000. *Mozambique: Rainy Seasons Bring New Threats*. http://www.reliefweb.int/IRIN/archive.phtml (August 3 2000)

Barry, Roger G. and Richard J. Chorley. 1998. *Atmosphere, Weather and Climate*. Seventh edition. London and New York: Routledge.

Batisse, Michel. 1996. Incertitude sur le Changement Climatique: Sortir de l'Ombre. Paris: *UNESCO/SOURCES*, n° 77.

Bolle, H. J. 1996. The Climate System and Global Change. In *Course on Climate Change Impact on Agriculture and Forestry*: Proceedings of the European School of Climatology and Natural Hazards Course held in Volterra, Italy, 16-23 March 1996. European Commission.

Boon, E. K and Hens, L. 1994. Sustainable Development in Practice. In *Environment and Development Education for Developing Countries*: Proceedings of a Seminar held in Accra, Ghana, from August 2 to 4, 1994. Brussels: Department of Human Ecology (VUB).

Cissé, Moussa, Youba Sokona and Jean P. Thomas. 1997. *Information Note on the Implementation in Africa of the United Nations Framework Convention on Climate Change*. Dakar: Enda Tiers monde. http://www.enda.sn/energie/accessener.htm (December 4 1999)

Cowie, Jonathan.1998. *Climate and Human Change: Disaster or Opportunity?* New York and London: The Parthenon Publishing Group.

Dessus, Benjamin and Michel Colombier. 1998. Développement durable: Réinventer la solidarité. *Courrier de la Planète/global Change*, Mars-Avril 1998: 33-34.

Franceys, R., J. Pickford and R. Reed. 1992. *A Guide to the Development of On-site Sanitation*. Geneva: World Health Organisation.

71

Goodall, Jane. 1991. What Effect Might Deforestation Have on Global Climate? In *Global Warming: The Debate*, edited by Peter Thompson. Chichester: John Wiley & Sons Ltd.

Harrison, Paul. 1987. *The Greening of Africa: Breaking through in the Battle for Land and Food*. London: Paladin Grafton Books.

Houghton, J. T. 1991. The Case for the Greenhouse Effect. In *Global Warming: The Debate*, edited by Peter Thompson. Chichester: John Wiley & Sons Ltd.

Hulme, M, D. Conway, P.M. Kelly, S. Subak and T. E. Down. 1995. *The Impacts of Climate Change on Africa*. Stockholm: Stockholm Environment Institute.

Intergovernmental Panel on Climate Change (IPCC). 1992. *Global Climate Change and the Rising Challenge of the Sea*. Report of the Coastal Zone Management Subgroup. The Hague (Netherlands).

Intergovernmental Panel on Climate Change (IPCC). 1995. *IPCC Second Assessment: Climate Change 1995*. A Report of the IPCC. Geneva.

Intergovernmental Panel on Climate Change (IPCC). 1998. *The Regional Impacts of Climate Change: An Assessment of Vulnerability,* edited by Robert T Watson, Marufu C. Zinyowera and Richard H. Moss. A Special Report of IPCC Working Group II. Cambridge: Cambridge University Press.

Justus, J. R. 2000. *Greenhouse Effect and Global Climate Change*. The National Council for Science and the Environment (NCSE). Washington.
http://www.cnie.org/nle/ebgccef.html (July 14 2000)

Kanté, Bakary. 1999. Le Developpement Durable, Seule Option Viable. *Liaison Energie-Francophonie* 43: 3-5.

Le Treut, H. 1997. Climate of the Future: An Evaluation of the Current Uncertainties. In *Climates and Societies: A climatological Perspective - A contribution to Global Change*, edited by M. Yoshino et al. Boston: Kluwer Academic Publishers.

Lewis, Peter, ed. 1998: *Africa: Dilemmas of Development and change*. Westview Press.

Mainguet, Monique. 1999: *Aridity: droughts and human development*. Berlin: Springer.

Michaels, P. J. 1991. Global Warming: Beyond the Popular Consensus. In *Global Warming: The Debate*, edited by Peter Thompson. Chichester: John Wiley & Sons Ltd.

Martin, P. H. 1996. Regional Aspects of Climate Change. In *Course on Climate Change Impact on Agriculture and Forestry*: Proceedings of the European School of Climatology and Natural Hazards Course held in Volterra, Italy, 16-23 March 1996. European Commission.

Ministère de l'Environnement et de la Protection de la Nature (MEPN). 1997. *Convention Cadre des Nations Unies sur les Changements Climatiques: Communication Initiale du Senegal.*Dakar: MEPN.

Ministère de l'Environnement et de la Protection de la Nature (MEPN). 1999. *Convention Cadre des Nations Unies sur les Changements Climatiques: Deuxième Communication du Senegal.*Dakar: MEPN.

Ministère de l'Environnement et de la Protection de la Nature (MEPN). 1999. *Stratégie Nationale Initiale de Mise en Oeuvre de la Convention Cadre des Nations Unies sur les Changements Climatiques.* Dakar: MEPN.

Mintzer, Irving M. 1992. *Confronting Climate Change: Risks, Implications and Responses.* Cambridge: Cambridge University Press.

Nilsson, Annika. 1992. *Greenhouse Earth.* Chichester: John Wiley & Sons.

Oberthur, Sebastian and Herman E. Ott. 1999. *The Kyoto Protocol: International Climate Policy for the 21st Century.* Springer-Verlag Berlin Heidelberg.

Ojo, O. 1997. Society-Climate Systems in Tropical Africa. In *Climates and Societies: A Climatological Perspective - A contribution to Global Change*, edited by M. Yoshino et al. Boston: Kluwer Academic Publishers.

Otieno, Dorothy. 1999. Climate Change: Africa's Nightmare. Climate Network Africa (CNA). http://www.meteo.go.ke/can/index.html (July 3 2000)

Ribot, Jesse C., Antonio Rocha Magalhaes and Stahis Panagides. 1996. *Climate Variability, Climate Change and Social Variability in the Semi-Arid Tropics.* Cambridge: Cambridge University Press. http://www.enda.sn/energie/cc/jesse.htm (December 4 1999)

Rosenzweig, Cynthia and Daniel Hillel. 1998. *Climate Change and the Global Harvest: Potential Impacts of the Greenhouse Effects on Agriculture.* New York: Oxford University Press.

Roqueplo, Philippe. 1993. *Climats sous surveillance: limites et conditions de l'expertise scientifique.* Paris: Economica.

Sokona, Youba, Stephen Humphreys and Jean P. Thomas. 1998. *Sustainable Development: A Centrepiece of the Kyoto Protocol - An African Perspective.* Dakar: ENDA Tiers monde.

http://www.enda.sn (December 4 1999)

Spore CTA, Information pour le Development Agricole des pays ACP. 1998. *Conflit International de l'eau douce.* Spore CTA n° 74.

Suliman, M. 1990. *Greenhouse Effect and its Impact on Africa.* London: Institute for African Alternatives.

United Nations Food and Agriculture Organisation (FAO). 1997. *Reformer les Politiques dans le domaine des Ressources en Eau: Guide des Méthodes, Processus et Pratiques* in Bulletin FAO d'Irrigation et de Drainage n° 52. Rome: FAO

United Nations Food and Agriculture Organisation (FAO). 1998. *Potential Impacts of Sea-Level Rise on Populations and Agriculture.* Rome: FAO

Wayne, A. Morrissey. 1998. *Global Climate Change: A concise history of negotiations and chronology of major activities preceding the 1992 UNFCCC.* Committee for the National Institute for the Environment. Science, Technology and Medicine Division. http://www.cnie.org/nle/clim-6.html (July 14 2000)

Whyte, I. D. 1995. *Climatic Change and Human Society.* London, New York, Sydney: Arnold.

Woodwell, Georges. M. 1992. Role of Forests in Climatic Change. In *Managing the World's Forests: Looking for balance between conservation and development,* edited by Narendra, P. Sharma. World Bank.

World Bank. 1996. *African Water Resources: Challenges and Opportunities for Sustainable Development.* World Bank Technical Paper n° 331: Africa Technical Department Series.

World Commission on Environment and Development (WCED). 1987. *Our Common Future.* Oxford, New York: Oxford University Press.

World Health Organisation (WHO). 1996. *Climate Change and Human Health.* Geneva: World Health Organisation.

World Health Organisation (WHO). 1997. Health and Environment in Sustainable Development: Five Years after the Earth Summit. Geneva: WHO

World Resources Institute (WRI), United Nations Environment Programme (UNEP), United Nations Development Programme (UNDP) and World Bank (WB). 1998. *World Resources: 1998 – 1999.* New York, Oxford: Oxford University Press.

World Wide Fund For Nature (WWF). 1993. *Sustainable Use of Natural Resources: Concepts, Issues, and Criteria.* A WWF International Position Paper. Gland, Switzerland: WWF.

Xoagub, François. 1997.Our Dams Can with El Nino. In the *Namibian*, October 17 1997. http://www.darwin.bio.uci.edu/~sustain/ENSO.html (July 17 2000)